GARDENING IN SUMMER-DRY CLIMATES

GARDENING IN SUMMER-DRY CLIMATES

PLANTS FOR A LUSH, WATER-CONSCIOUS LANDSCAPE

NORA HARLOW AND SAXON HOLT

Frontispiece: Plants adapted to summer-dry climates are displayed in a formal arrangement astride a shallow channel at The Huntington Botanical Gardens in Los Angeles.

All photos are by Saxon Holt except for the following: page 126 (bottom) by Julie Ann Kierstead, page 203 (bottom) by Arvind Kumar, Wikimedia Commons, used under a CC Attribution-by Share Alike 3.0 Unported license, and page 259 (left) by Margo C. Bors.

Published in 2020 by Timber Press, Inc.
The Haseltine Building
133 S.W. Second Avenue, Suite 450
Portland, Oregon 97204-3527
timberpress.com

Printed in China
Text and cover design by Anne Kenady Smith

ISBN 978-1-60469-912-8
Catalog records for this book are available from the Library of Congress and the British Library.

CONTENTS

PREFACE

It is an exciting time to be a gardener. We all understand the importance of gardening within our planet's means, that Garden Earth requires our stewardship, and that we gardeners can be a positive force for sustainability.

We began planning this book shortly after finishing *Plants and Landscapes for Summer-Dry Climates of the San Francisco Bay Region*, published in 2004 by the East Bay Municipal Utility District. That book, though widely acclaimed, targeted a narrow geographic zone. We believe that many more gardeners can benefit from what we all have learned about climate change in recent years.

In summer-dry climates, rainless summers are not drought; they are normal. Gardening in this climate, where water is so precious, requires a careful choice of plants that can thrive in dry summers and, in that seasonal corollary, wet winters.

This is not a no-summer-water book. Some plants will suffer if watered in the summer, but most fare better with at least some water and most gardeners want some degree of lushness around their homes. Mindful gardeners can use supplemental irrigation efficiently to support resilient landscapes that benefit all creatures that inhabit them, from human to mycorrhizal.

Low water use allows for vastly different types of gardens across the wide geographic zone we include here, from British Columbia to Baja. Each gardener should understand that a low-water plant in Seattle might require a lot of water in San Diego, and a low-water plant in Los Angeles might rot in a wet Portland winter.

Successful, low-impact, beautiful gardens are an art form that begins with acknowledging the natural landscape that surrounds the garden and orchestrating seasonally varied plants, large and small, that are climate-adapted and in harmony with their surroundings. Plants and landscapes from summer-dry climates all over the world can help us create our own plant tapestries and garden symphonies.

The plants we showcase in this book are not intended in any way to limit your choices. We hope they will inspire you to look beyond what is described here. There are many more selections, in almost every genus, that we didn't have space to include. Plant collectors and nursery professionals continue to develop and introduce new plants, seemingly almost every day.

Acknowledgments

Our most difficult task in preparing this book was narrowing down the plant list, and for this we had extraordinary help from some of the very best experts in the world. No doubt we will fail to mention every individual who influenced our thinking or turned our attention to summer-dry plants we had not considered.

In today's pantheon of West Coast garden authorities we are deeply indebted to Paul Bonine of Xera Plants in Portland; Carol Bornstein of the Natural History Museum of Los Angeles County; Emily Griswold from the University of California, Davis, Arboretum; and Richard Turner, editor emeritus of *Pacific Horticulture* magazine for peer review of the plant list. These four experts, broadly representing the major climate zones across the summer-dry Pacific Coast, not only reviewed the list plant by plant, but contributed authoritative knowledge and experience on plant culture and water use.

Before we narrowed down the list, we received invaluable suggestions from these and other plant and garden luminaries: Stewart Winchester, Marilee Kuhlmann, Kathy Musial, Roger Raiche, Warren Roberts, Bart O'Brien, and Sean Hogan.

Many of the gardeners and designers who allowed us to photograph their gardens added real-life suggestions, and for this we wish to thank Charlotte Torgovitsky, Michelle Derviss, Mary and Lew Reid, Roger Greenberg, Greg Shepherd, John Kuzma, Mike Smith, Shelagh Tucker, Stacie Crooks, Richie Steffen, John Albers, Ernie and Marietta O'Byrne, David Feix, Patrick Anderson, Susan Gottleib, Jim Bishop, Scott Borden, John Greenlee, Dave Buchanan, Nan Sterman, Debra Lee Baldwin, David Fross, Urban Water Group, Arcadia Studio, Puck Erickson, Ground Studio, Brian Kemble, Jo O'Connell, John Gabbert, Susan Schaff, Janet Sluis, Diana Magor, Jenn Simmons, Ruskin Gardens, Lili Singer, Gary Ratway, Nancy Roche, Judy Adler, Kathy Kramer, Michael Thilgen, Kate Frey, and Billy Krimmel.

The following public gardens allowed us special access for photography, and we are indebted to them also for the time they spent confirming plant identifications. Each is a valuable resource to gardeners in summer-dry climates: San Diego Botanic Garden, Natural History Museum of Los Angeles County, Los Angeles County Arboretum and Botanic Garden, The Huntington Botanical Gardens, Leaning Pine Arboretum, Santa Barbara Botanic Garden, Regional Parks Botanic Garden, Blake Garden at UC Berkeley, The Ruth Bancroft Garden and Nursery, University of California Botanical Garden at Berkeley, San Francisco Botanical Garden, UC Davis Arboretum, Elisabeth Carey Miller Botanical Garden, Bellevue Botanical Garden, University of Washington Botanic Gardens, The Annenberg Retreat at Sunnylands, The Living Desert Zoo and Gardens, and Denver Botanic Gardens.

We are grateful as well for the calm, competent, and always thoughtful guidance and assistance of Timber Press editor-in-chief Tom Fischer and for the individual and collective contributions of the entire Timber Press team.

Finally, for inspiration we visited many parks and open spaces throughout the West. There is no better way to learn how to garden than to visit natural areas. Thanks, Flora!

SAXON HOLT AND NORA HARLOW

A tapestry of heaths and heathers, barberries, and dwarf conifers creates a richly textured Seattle garden.

Gardening Where You Are

GARDENERS IN SUMMER-DRY climates don't need charts and maps or statistics on rainfall to tell them that their climate is summer-dry. Our climates may be called hot- or cool-summer mediterranean, oceanic, semi-arid, submediterranean, or west coast maritime, but we are all summer-dry and we have much in common as gardeners.

Not that there are no regional garden styles. Gardens reflect the convergence of the physical and the social—of the physical facts of climate, topography, and natural vegetation and the personal and collective experience of those who live and garden there.

Regional variations are seen all along North America's summer-dry West Coast. Gardens in Seattle or Vancouver are distinctly of the Pacific Northwest. For the most part, they little resemble the gardens of Los Angeles. Even when we try to replicate a garden from another place, say a California garden in Portland or a Northwest garden in San Francisco, we can hardly avoid expressing something of the local vernacular.

< A sinuous stairway invites exploration up and into a San Diego garden of succulents and palms.

There are measurable physical differences from north to south as well. Not just in rainfall totals but in the timing and intensity of rain. Not only summer drought but how long it lasts and how hot and dry the soil and air. Coastal Oregon and California's Central Valley are palpably different kinds of summer-dry, not only in the number of cloudless days but in the brightness of the sun, the quality of the light, and how colors present themselves in the presence or absence of mist or fog.

The similarities among summer-dry climates are every bit as striking as their differences. We plant mostly in fall, less often in spring as most of the world does. In summer, many of our finest perennials go from full-on floriferous to a restful dormancy. We can grow plants from just about anywhere in the world, but if those plants are accustomed to summer rainfall, we will need to water them.

With a warming climate and a growing population, our challenges are becoming increasingly alike as well. From north to south, when winter rain and snow fall short we can expect summertime restrictions on garden watering. North to south, wildfires are ever larger, more destructive, and harder to control. Invasive species and the near-catastrophic loss of natural wildlands are challenges we share with one another and with the world.

Gardeners up and down the Pacific coast also share an upbeat conviction that the way we garden can make a difference. In the face of worldwide habitat loss, species extinction, and unsustainable pressures on natural resources, we have moved decisively to reduce our impact on water supplies, to make gardens that attract and sustain wildlife, to use and reuse local materials, and to work with rather than fight the summer-dry climate.

The summer-dry garden is not necessarily a dry garden. Plants from other summer-dry climates may need occasional or even moderate watering to make it through our particular version of summer-dry, especially if winter rainfall has been less than normal. A plant can be well adapted to summer dryness in its natural setting and still need a little help from the gardener.

Gardening in harmony with the summer-dry climate begins with an understanding of where, exactly, you are. It is also useful to know something about the conditions in which the plants you select grow naturally.

Summers are dry or mostly dry, but how dry and for how long? Winters are wet or usually wet, but do storms come rarely with heavy rain, or are there many winter days with only a light drizzle? Summer heat and winter cold both help to determine what plants we can successfully grow. Do winter temperatures drop below freezing? How far below, how often, and for how long?

< A Seattle garden features rosemary, lavender, sage, and other sun-loving plants common in California gardens.

^ Perennials are artfully displayed at the University of Washington Botanic Gardens in Seattle, where precipitation is generally light but frequent, giving the impression of a much rainier climate than its annual average of 38 inches.

^ Mountains and valleys north of Santa Barbara run west to east, instead of the more typical north–south orientation, funneling summer fog and cooling ocean air inland toward the Presqu'ile Winery.

THE SUMMER-DRY CLIMATE

< *Agave americana* and palo verde (*Parkinsonia* 'Desert Museum') are well suited to the climate of the Natural History Museum of Los Angeles County; annual rainfall in the region is lower than in many other summer-dry climates of the world.

SUMMER-DRY CLIMATES OF THE WORLD TEND to be on or near the west coasts of continents where temperatures are moderated by proximity to the ocean and where semi-permanent atmospheric pressure systems influence the path of approaching storms. In these regions, summers may be cool, warm, or hot; winters are mild; and rainfall is concentrated in winter.

The seasonality of precipitation is the defining characteristic of summer-dry climates, which otherwise vary widely in how rainy days are spread out over the year. Rain falls mostly in winter in both Cape Town, South Africa, and Los Angeles, California, but in a year of normal rainfall, Cape Town receives 2 inches of its 20-inch annual total, or 10 percent, in summer. Los Angeles receives 0.14 inch, or less than 1 percent, of its 15-inch total in summer months. Los Angeles has a more intensely summer-dry climate.

Even within the same summer-dry region of the world, climates differ markedly in the amount of rain that falls in the mostly summer-dry season. In the region surrounding the Mediterranean Sea, the northwest coast tends to receive some summer rain. Some parts of that coast—near Barcelona, for example—may see so much of their annual rainfall in summer that they can hardly be called summer-dry. The southern Mediterranean coast, from Morocco to Tunisia, and the eastern Mediterranean, from southwestern Turkey to Israel, usually are much drier than the north coast, with little or no rain in summer.

In any summer-dry climate, annual rainfall and precipitation patterns can vary dramatically from one year to another. Abnormally dry or wet winters may repeat for several to many years. A series of dry years may be followed by an exceptionally wet year or by a year in which the rainy season is unusually long or short. Variability is one of the few constant features of the summer-dry climate.

Spiky agaves, aloes, and a succulent euphorbia flourish in San Diego, where average annual rainfall is 10 to 12 inches.

SUMMER-DRY CLIMATES OF THE PACIFIC COAST

THE WEST COAST OF NORTH AMERICA RECEIVES most of its rain in winter, with the summer season mostly to almost completely dry. From Vancouver to San Diego, there typically are weeks or months in the warm season when no rain falls.

There is great variation, of course, from north to south in rainfall amounts and in timing of the wet and dry seasons. To the north, the length and intensity of the warm-season drought decreases, with more rain falling in spring, summer, or early fall. Seattle receives an annual average of 38 inches of rain, with about 12 percent falling in the summer months and rain spread out over about 150 days. To the south, the dry period lengthens and increases in intensity. San Diego receives an average of 10 to 12 inches of rain, with less than 2 percent falling in summer and only 43 days with any measurable rain at all.

^ A walled patio garden in Los Altos, one of the San Francisco Bay Area's many microclimates, is sheltered from the ocean by the coastal mountains yet open to cooling summer breezes off the bay.

Temperatures also trend higher from north to south. The annual average high in Los Angeles is 15 to 20 degrees above that in Vancouver, and the average low is at least 10 degrees higher. Averages mask the extremes, and extremes matter. Summertime highs in Los Angeles often edge into the 100s for days at a time. In Vancouver, the hottest summer days usually are in the mid-80s. Lows at or below freezing in Vancouver occur an average of 14–40 nights a year. Los Angeles rarely sees even a light frost.

Differences in local climates or microclimates can be even more remarkable. Summer temperatures in Burbank, California, may be 10 degrees hotter than temperatures 9 miles south in downtown Los Angeles. Downtown Los Angeles may be 10 degrees hotter than Santa Monica, less than 15 miles away. Temperatures farther inland can be 20 to 30 degrees hotter than those right along the coast.

Neighborhoods on the west side of San Francisco often are blanketed with thick summer fog, while those just a few blocks east are basking under sunny skies. When dense fog spreads across the bay to envelop Oakland, a short trip through the tunnel under the Oakland hills usually delivers bright sunshine in a little over half a mile.

Similar variations are found to the north. Fifty miles northwest of Seattle, the city of Sequim receives less than half as much rain. Olympia, almost 50 miles to the southwest, receives a foot more rain and has a frost-free growing season more than 80 days shorter.

^ Perennials in the Waterwise Garden of Bellevue Botanical Garden near Seattle thrive in the mild, moist climate of Puget Sound.

^ Summer fog brings significant moisture and cool temperatures to the San Francisco Botanical Garden.

ORIGINS OF THE PACIFIC COAST CLIMATE

THE CIRCUMSTANCES THAT PRODUCE THIS distinctive arrangement of climates are complex and interrelated. Summer-dry climates, as with climates worldwide, are largely determined by such globally significant factors as distance from the equator, prevailing winds, the direction and temperature of ocean currents, and the shifting location and strength of atmospheric pressure systems that govern the path of approaching storms. Weather along the Pacific coast is determined as well by the meandering jet stream, by fluctuating ocean temperatures associated with El Niño and La Niña events, and by the size, timing, and moisture content of atmospheric "rivers" that make landfall along the coast. Regional and local variations in climate are shaped by distance from the ocean and topography.

Proximity to the ocean plays a central role. Oceans gain and lose heat more slowly than adjacent land. As a result, air moving over the water is warmer in winter and cooler in summer than temperatures on land. This has a moderating effect on climates along the Pacific coast, an effect that diminishes as you move inland. Proximity to the ocean also brings fairly reliable daily breezes and sometimes strong winter winds.

Air temperatures are further modulated by the temperature and direction of ocean currents. From about the border between British Columbia and Washington, the cold California Current flows south along the coast and the warmer Alaska Current flows north. In summer, as the California Current flows south, winds from the northwest propel the surface water south and then west, causing even colder water to rise from below to replace it. This creates a band of especially cold water along the coast, where moisture condenses and fog forms as warmer winds from the ocean pass over it. As the land heats up later in the day, the fog usually dissipates, but it often stays long enough for fog drip from trees to soak the ground.

Precipitation patterns on the west coast of North America are largely determined by the location and strength of two seasonally shifting atmospheric pressure systems over the northern Pacific Ocean. In winter a low-pressure trough over the Gulf of Alaska, the Aleutian Low, moves south, sending storms toward much of the Pacific coast. In summer the low-pressure trough weakens and moves north and a high-pressure ridge, the North Pacific High, takes its place. The high-pressure ridge sends storms northward around it, blocking storms from more southerly parts of the coast. Sometimes this ridge stalls off the coast for weeks or months at a time, keeping most of the Pacific coast dry.

Once storms reach the coast, topography ensures that their effects are not uniformly felt. Mountains exert a particularly large effect on precipitation, especially if oriented perpendicular to prevailing winds. When winds reach the mountains, air rises and cools, clouds form, and water vapor condenses and precipitates. As the cooled air descends to lower elevations on the other side, the air warms and the wind is drier. The higher the mountains, the greater the effect, but air masses can be blocked or redirected by fairly low mountains, and cold air can pool even in small depressions.

Elevation and latitude have similar effects on temperature, which decreases fairly consistently from south to north and from lower to higher elevations. Moving north 300 miles is, on average, equivalent to an increase of 1,000 feet in elevation. Precipitation usually increases with elevation, but the rate of increase depends on factors that affect the amount of moisture in the air, such as distance from the ocean or other large bodies of water.

Aspect also affects climate. In the northern hemisphere, south-facing slopes usually are warmer and drier than north-facing slopes. South-facing slopes receive direct sunlight and both temperature and evaporation usually are higher. East-facing slopes receive morning sun while west-facing slopes receive hotter afternoon sun. Vegetation that thrives on south-facing slopes at higher elevations or in northern latitudes may grow on north-facing slopes at lower elevations or in more southerly latitudes. Vegetation differs with aspect even at similar latitudes and elevations.

^ The Ruth Bancroft Garden and Nursery in Walnut Creek, California, protects and displays many plants from arid and semi-arid parts of the world.

TOPOGRAPHY AND CLIMATE

BECAUSE OF THE NORTH–SOUTH ORIENTATION of the mountains and their effects on temperature and precipitation, local climates on the Pacific coast of North America tend to be more similar north to south than east to west as is common elsewhere in temperate parts of the world. Changes in climate are fairly gradual moving north to south but can be extreme moving west to east.

From Alaska to Baja California, the north-south trending mountain ranges interrupt, deflect, or otherwise channel the prevailing westerly winds. From the Vancouver Island Ranges and the Olympic Mountains south through the Coast Ranges of Washington, Oregon, and California, the effect is the same: cooler summers, warmer winters, higher humidity, and more precipitation along the coast and on west-facing slopes than even a few miles inland.

East of the coastal mountains is a series of usually north-south trending valleys. The Fraser Valley in British Columbia, the Willamette Valley in Oregon, and the Central Valley in California are the major valleys, but there are more. Most of the valleys are in the rain shadow of the coastal mountains, especially close to the east-facing slopes. Climates in the valleys are warmer in summer, colder in winter, and lower in precipitation than along the coast. Temperatures also vary more widely from day to night.

Farther inland, on the eastern side of the valleys, the Cascade Range and the Sierra Nevada form a second north-south chain from British Columbia to just north of the Los Angeles Basin. These mountains are higher, with peaks in the Cascades rising to well over 10,000 feet and more than 100 peaks over 13,000 feet in the Sierra Nevada. Beyond these ranges are the harsher, more continental climates of the Great Basin and the Rocky Mountains, where winters are cold, summers are hot, and what precipitation there is often falls as snow.

The coastal and inland mountain ranges, although continuous, are not without gaps through which weather patterns interact. From north to south, several major rivers cut through the mountains to reach the ocean, creating passages for ocean breezes to flow inland in summer and cold northerly winds to bring winter storm systems to the coastal edge. Such passages are formed by the Fraser River in British Columbia, the Columbia River in Washington and Oregon, and the Sacramento River in California. Dozens of smaller rivers, streams, and mountain passes also funnel winds and weather in both directions.

The mountains contribute to the temperature inversions so common in west coast cities that are surrounded by mountains. Cold air sometimes flows down from mountain peaks, pushing under warmer air rising from the valley or the coastal plain. The warm air above creates a cap that prevents the mixing of cold and warm air and traps dust and smog beneath it. Los Angeles is famous for its temperature inversions, but inversions also occur from Seattle to Olympia in Puget Sound, from Portland to Eugene in the Willamette Valley, and from Redding to Bakersfield in California's Central Valley.

Another weather-related phenomenon brought about in part by topography is inherent in parts of the Pacific Coast, especially to the south. Hot, dry, easterly winds come off the inland valleys and deserts in fall, bringing sudden increases in temperature and drops in humidity to the coast. Called Diablos in north-central California and Santa Anas in Los Angeles, these unusually strong, downslope winds fan wildfires into flaming infernos.

A similar phenomenon, called the Brookings effect, occasionally occurs in southwestern Oregon, where northeasterly winds pour down the Chetco River toward the coast at Brookings, lowering humidity and raising temperatures dramatically.

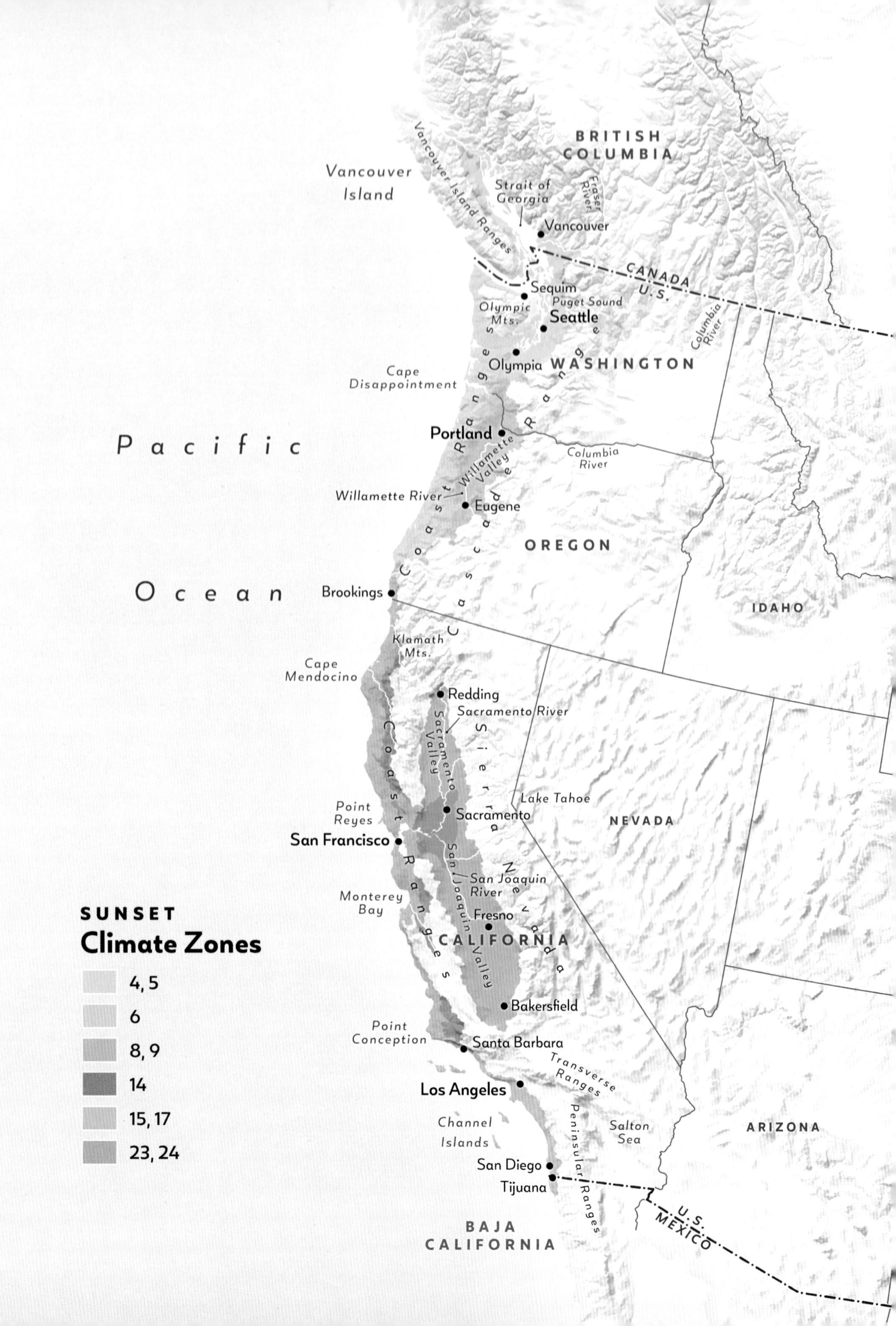

BRITISH COLUMBIA
Vancouver Island
Vancouver Island Ranges
Strait of Georgia
Fraser River
Vancouver
CANADA
U.S.
Sequim
Puget Sound
Olympic Mts.
Seattle
Columbia River
Olympia
WASHINGTON
Cape Disappointment
Cascade Range
Coast Range
Portland
Willamette Valley
Columbia River
Pacific Ocean
Willamette River
Eugene
OREGON
Brookings
IDAHO
Klamath Mts.
Cape Mendocino
Redding
Sacramento River
Sacramento Valley
Sierra Nevada
Coast Ranges
Lake Tahoe
NEVADA
Point Reyes
Sacramento
San Francisco
San Joaquin River
San Joaquin Valley
Monterey Bay
Fresno
CALIFORNIA
SUNSET
Climate Zones
4, 5
6
8, 9
14
15, 17
23, 24
Bakersfield
Point Conception
Santa Barbara
Transverse Ranges
Los Angeles
Peninsular Ranges
Channel Islands
Salton Sea
ARIZONA
San Diego
Tijuana
U.S.
MEXICO
BAJA CALIFORNIA

^ Oaks and California buckeyes stand out prominently against the golden grasses of summer on Mount Burdell, a 1600-acre open space preserve 30 miles north of San Francisco.

Lush green grasses and budding oaks are highlights of spring in coastal mountains near Gilroy, California, 30 miles south of San Jose. >

Climate Zones and Vegetation

There are gradual but unmistakable shifts in climate as you travel north to south along North America's Pacific coast. The climate effects of latitude, topography, and atmospheric pressure systems are prominently reflected in the natural vegetation.

Climate zones developed for the *Sunset Western Garden Book* include six zones that depict the moderating influence of the ocean along the coast. Three zones reflect the somewhat more extreme climates of the inland valleys. There are more than a dozen other climate zones within this topographically complex part of the world, but these few capture the essential traits of some of the Pacific coast's summer-dry climates.

Sunset zones 5, 17, and 24 represent the mild climates of a coastal plain that extends in a narrowing and widening strip along the coast from southeastern Vancouver Island to Tijuana, Mexico. Inland and adjacent to the coastal plain, zones 4, 15, and 23 represent a coastal climate affected slightly less by the ocean and slightly more by the continental weather systems of the interior. Zones 6, 8, and 9 are inland valley climates, hotter in summer and colder in winter than along the coast.

Coastal Pacific Northwest

Zones 5 and 4, the most northerly pair of the six coastal zones, extend from southwestern British Columbia to southwestern Oregon and include some of the wettest climates in North America. Average rainfall ranges from 60 to more than 150 inches a year. The climate supports a mosaic of habitats, from tidal mudflats and salt marsh estuaries to grassy headlands, sand dunes, and substantial remnants of wet Sitka spruce forest with an understory of Oregon grape, salal, and ferns.

Summers in these north coast zones are cool and windy but mostly dry. Even in the wettest forests it may not rain for several weeks. Summertime moisture is regularly supplied by night and morning fogs that may or may not lift by afternoon. Winters are wet and cool with frequent clouds but only occasional frosts. Humidity is high year-round.

Coastal Northern California

To the south of zones 5 and 4, Sunset zones 17 and 15 run from a few miles north of the Oregon–California border south to Point Conception, where the atypically west–east trending mountains of the Transverse Ranges separate northern from southern California. Average rainfall varies, from 77 inches in Brookings, Oregon, to 23 inches in San Francisco and 16 inches near Point Conception. Along the coast, summers are cool with reliably foggy mornings. Slightly inland, summers are warm to hot and dry. Winters both along the coast and inland are mild with alternating sunny and rainy days.

Vegetation in zones 17 and 15 shifts from fog-drenched redwood and Douglas-fir forest in the north to dry-adapted coastal scrub and chaparral in the south, with a complex array of plant communities in between. In the hill-and-valley terrain around San Francisco Bay, it is not unusual to find tidal marshes, grassland, coastal scrub, chaparral, oak woodland, and redwood or mixed-evergreen forest within a few miles of each other.

Coastal Southern California

Sunset zones 24 and 23 run down the coast from Point Conception and Santa Barbara to

^ On rocky headlands overlooking the ocean 15 miles north of San Diego, Torrey Pines State Natural Reserve protects several thousand of the rarest North American pines (*Pinus torreyana*) and one of the last large salt marshes and waterfowl refuges in southern California.

San Diego and Tijuana. These coastal zones are narrow where mountains are closest to the sea and wider where broad, flat basins form alluvial plains with mountains some distance back from the water. Rainfall totals range from less than 20 inches in Santa Barbara to 12–15 inches in the Los Angeles Basin and 10–12 inches in San Diego. Summers are cool to warm along the coast and mostly hot inland. Winters are mild, but slightly inland, winter temperatures in the 80s or low 90s are not uncommon.

Much of this southern coastal region is highly urbanized, and the nearby mountains and beaches are heavily used for recreation. Remnants of diverse habitats still survive in protected areas, including dunes and bluffs, estuaries and wetlands, riparian corridors, wooded canyons, and chaparral-covered slopes. The varied terrain on which development has occurred surrounds many residential areas with undeveloped urban wildlands.

^ Garry oak and madrone provide light shade for spring wildflowers in the Camassia Natural Area, a 26-acre Nature Conservancy preserve south of Portland overlooking the Willamette River.

Inland Valleys

Sunset zones for the major inland valleys are zone 6 for the Willamette Valley and zones 8 and 9 for the floor and foothills of the Central Valley. The valleys are lowlands surrounded by mountains on all sides, with rain shadow effects from mountains to the west and gaps through which climate is affected by both maritime and continental winds and weather systems.

The lowlands around Puget Sound and the Strait of Georgia are similarly affected by surrounding mountains and passes through them, but the overwhelming influence of exposure to the ocean places this lowland region, including Vancouver and Seattle, in coastal Sunset zones 5 and 4.

Willamette Valley

The Willamette Valley extends from just north of Portland to just south of Eugene, a distance of about 150 miles. The Willamette River runs the length of the valley, from the Douglas-fir forests in the mountains above Eugene to the Columbia River less than 10 miles northwest of Portland. Annual rainfall ranges from 36 inches in Portland to 46 in Eugene and 60–80 in parts of the adjacent foothills. Summers are warm to hot and mostly dry. Winters are mostly cool and cloudy, but Arctic air flowing down through the Columbia River Gorge occasionally brings frigid temperatures to much of the valley.

Historically the Willamette Valley hosted seasonally wet meadows and riparian woodlands in the valley floodplains, Garry oak woodlands in the drier western foothills, and Douglas-fir forests in the eastern foothills. As elsewhere up and down the Pacific coast, the relatively flat, fertile floodplains proved irresistible for farming and were drained and planted long ago. Much of the rest of the valley has been developed, but vestiges of the natural vegetation remain.

^ The semi-arid Carrizo Plain, just west of the southern San Joaquin Valley, hosts impressive displays of spring wildflowers in years of heavy rainfall.

Central Valley

California's Central Valley consists of the northern Sacramento Valley, drained by the south-flowing Sacramento River, and the southern San Joaquin Valley, drained in part by the north-flowing San Joaquin River. The two rivers join at the 750-square-mile Sacramento–San Joaquin River Delta before flowing into San Francisco Bay.

The direct connection through the delta to the bay and the ocean places the central portion of the valley, including Sacramento, in Sunset zone 14, which is otherwise reserved for intermountain valleys closer to the coast.

The climate of the Central Valley is hot summer-dry, shading into semi-arid or arid to the south. Winters are mostly cool and cloudy but with many sunny days and sometimes heavy rain. Dense ground fog often forms in winter as cold air drains off the mountains and pools in the valley, causing temperature inversions that can last for weeks. Rainfall totals drop from north to south, with Redding receiving 35 inches, Sacramento about 18, Fresno 11–12, and less than 6 inches in Bakersfield.

Once covered in grassland and oak woodland to the north, wetlands in the delta, and grassland, seasonal wetlands, and scrublands to the south, the Central Valley is now largely urban or suburban and agricultural. Little of the natural vegetation remains except in protected ecological reserves and wildlife refuges. With irrigation, the Central Valley is enormously productive, providing dozens of crops to nationwide and worldwide markets.

PLANTS FOR SUMMER-DRY CLIMATES

WHERE A PLANT GROWS NATURALLY IS NOT THE only or even the predominant fact that determines whether it will succeed in a particular garden. But native range or origin is a place to begin, and it is hard to believe that it is ever completely ignored.

Plants are advertised as "drought tolerant" or "California native" as if that alone could suggest their cultural preferences. Drought tolerant where? Native to what? A plant that needs no summer water along the northern California coast may do just as well without supplemental water in Seattle. That same plant may struggle in Sacramento even with regular summer water and die in Bakersfield with ample water and in shade.

Summer-dry parts of the world from which many of our garden plants derive are as varied in climate as our own. What do we learn from a label that says a plant is native to Australia? Most of Australia is arid or semi-arid or rainfall is distributed throughout the year. Summer-dry, winter-wet climates occur along the southwestern coast of Western Australia and parts of the southern coast of South Australia. Plants from wetter parts of southeastern Australia may need some summer water in Los Angeles but thrive in San Francisco on their own.

South Africa is another rich source of plants grown in summer-dry climates. Climates in South Africa range from arid in the northwest to year-round rainfall in the southeast, summer rainfall in the northeast, and summer-dry in the southwest. Climates of South Africa's summer-dry region, Western Cape Province, are drier moving north from the southwest corner near Cape Town and wetter moving east. Plants grown in North America's summer-dry climates come from both wetter and drier parts of Australia.

Even a plant native to the dry American Southwest may be accustomed either to little water year-round or to late-summer downpours from the North American monsoon. Parts of New Mexico, Arizona, and northern Mexico can receive half their yearly rainfall from summer monsoons, which only occasionally reach the deserts of California. Plants from many parts of the Southwest are marketed simply as drought-tolerant plants with no mention of their likely preference for a little summer water.

Rainfall is not the only variable affecting plants. The summer-dry Mediterranean region hosts an enormous variety of habitats and plant communities and is home to about 25,000 species, almost half of which are native nowhere else. The rugged and often steep terrain and the wide range of soils and rock types have resulted in many specialized environments to which plants have adapted and in which they have evolved. Plants grown in summer-dry climates along our Pacific coast come from many different parts of the region around the Mediterranean Sea.

^ *Yucca rostrata*, native to southwestern Texas and northern Mexico, is quite content in this Portland garden.

It is clear that not all plants described as "drought tolerant" or "Mediterranean" will be content in the same conditions. In your garden, some will need some summer water and some will need a little shade. A few will accept soggy soils in winter, but many, perhaps most, will not. Some will need more winter chill or summer heat than your garden can provide. Some will do well wherever you plant them, while others may refuse to settle down no matter what spot you try.

With so many choices and so many details to consider, how are we to know which plants we can successfully grow? For gardeners along North America's west coast, it makes sense to start with some of the many plants native to or commonly grown in our own local area or in similar parts of the floristically rich Pacific Coast region.

Manzanitas, ceanothus, coffeeberry, currants and gooseberries, mahonias, monkeyflower, mock orange, silktassel, toyon, wax myrtle, oceanspray, and elderberry are just a few of the native shrubs we share. Some of our native trees are madrone, buckeye, incense cedar, mountain mahogany, and many oaks and pines. Native perennials we have in common include penstemons, achillea, buckwheats, milkweeds, heucheras, stonecrops, irises, dudleyas, and more.

We also grow many of the same plants native to other parts of the world, and many more could be tried. You may see more conifers and maples, more heaths and heathers in Northwest gardens, and California gardens tend to have more succulents of all kinds. But many plants are grown in gardens throughout our summer-dry region: lavenders, rosemaries, artemisias, grevilleas, rockroses, eryngiums, euphorbias, agaves, and ornamental bunchgrasses are everywhere.

Not all perform equally well everywhere, of course. Some plants that flourish in southern California won't last a full year in colder and wetter parts of the Pacific Northwest. But many plants are widely adaptable, either because the species is genetically suited to varying soils and climates or because different populations of the species have their own specialized adaptations and requirements.

Growers are constantly breeding, selecting, testing, and introducing new varieties. Often the traits selected for are purely aesthetic—larger and more colorful flowers, smaller and more compact plants, longer periods of bloom, or repeat bloom. But hardiness and disease resistance are also high on the list. Growers have long been pushing the limits—from eucalypts that thrive in Portland to crape myrtles that succeed along the foggy coast.

Even some plants from summer-rainfall climates can be grown successfully here with moderate to occasional summer water if other conditions are favorable. Camellias, from summer-rainfall southern and eastern China and Japan, are on no one's list of drought-tolerant plants. With part shade and in moisture-retentive soils, mature camellias may be content with moderate summer water in gardens from Vancouver to Los Angeles.

^ In a Mendocino, California, garden, summer morning fog obscures the ocean just beyond the gate.

Most gardeners routinely consider climate, soils, and cultural preferences in selecting plants. Other important questions are less often addressed early enough or in sufficient detail to avoid later disappointment or regret. These, too, should influence the selection of plants for our gardens.

How much water do you want to devote to the garden? How much time? Can you embrace the browns and grays of summer dormancy and, if so, where and to what extent? Are you willing to let plants assume their full height and width? Compete for space and move about the garden? If so, how much vegetative autonomy can you accept?

More sensitive questions facing gardeners today have to do with plant origins or nativity. Do you want a garden of native plants and, if so, will these be exclusively natives or natives combined with plants from other parts of the world? What does "native" mean to you? Native to your local watershed? The shaded, north-facing slopes of your watershed? Native to your county? Your state? Your ecoregion?

There are no right or wrong answers to questions such as these. Both the questions and the answers are inherently personal. Your responses will help you seek out and select plants that suit your own needs and preferences. In almost every case you can choose a middle path.

A dry creek bioswale captures and absorbs stormwater runoff from roof and patio, helping to reduce the load on stormwater systems and replenishing groundwater supplies.

Some Thoughts on Design

THE IDEA OF GARDENING with plants adapted to climate is not a new one. Nor is it news, to gardeners at least, that gardens can serve purposes that go far beyond the aesthetically pleasing and decorative.

New today are the high expectations that are coming to be placed on designed landscapes to help solve some of the world's most daunting problems: a warming climate, the worldwide loss of biodiversity, and the disruption of natural systems that comes with development to serve a growing population.

Directly and indirectly, gardens can increase biodiversity. Gardens can be designed to filter and absorb stormwater runoff, keeping pollutants out of local waterways and helping to replenish groundwater supplies. Gardens can capture atmospheric carbon for storage in the soil, helping to reduce greenhouse gases that contribute to global warming.

At the same time, gardens in summer-dry climates must address such fundamental design issues as summer drought, waterlogged winter soils, and where to place the barbecue. Our landscapes are used for many purposes, prominently including relaxation, children's play, and social gatherings. Plantings are expected to frame or block views, provide shade, shelter the site from wind, and support pollinators, all while avoiding the use of invasive plants and protecting our homes from wildfire.

Solutions for one challenge can also be solutions for others. Managing stormwater on site can provide a supplemental water supply for the dry season while accommodating the wintertime drainage needs of dry-adapted plants. Siting plants away from the house for fire safety can provide convenient spaces for outdoor recreation and entertainment. Avoiding the use of pesticides and herbicides can promote carbon capture while creating a safe environment for birds and butterflies.

Perhaps the most revolutionary idea is also the most obvious to experienced gardeners. Garden design is not something you do once and then walk away, devoting subsequent months and years to the mostly futile task of maintaining the garden as designed. What brings any garden to life is the conscious decision to let go a little, to step back and let nature assert itself, welcoming change and a bit of wildness. In such a garden, the act of design is continuous and interactive, an ongoing exchange between gardener and garden.

^ A patio that invites opportunities for relaxing, entertaining, and enjoying an extended view is surrounded by succulents and low shrubs planted away from the home to create a firesafe landscape.

Perennials and shrubs at the Ruth Risdon Storer Garden at the UC Davis Arboretum were selected for their suitability for Central Valley gardens.

SUMMER-DRY AND WINTER-WET

LANDSCAPES IN SUMMER-DRY CLIMATES MUST be designed to deal with little or no rain for weeks or months in summer as well as sometimes heavy winter rains. Climate trends suggest that extended summer droughts and heavy winter rains may both become more common as global temperatures rise.

Climate-adapted plants and unwatered spaces address the fact of rainless summers. Saturated soils can be managed by adding amendments, planting on earth mounds or in raised beds, and guiding runoff away from plants that need good drainage. Rainwater harvesting can help with both winter-wet and summer-dry.

Climate-Adapted Plants

Not all plants marketed as "drought tolerant" or "low water" are well adapted to summer-dry climates. Summer-dry does not mean desert, where plants are adapted to brief and sudden summer downpours but cannot deal with heavy winter rains. Summer-dry does not mean mostly dry, high-elevation climates with frigid winters, where low-water plants from milder climates may not be low-water at all.

Plants from summer-dry climates are adapted to mild, wet winters and some degree of summer drought, but they vary in the amount of summer water they prefer. Some plants tolerate drought but don't prefer it. Others accept summer dryness only with afternoon shade. Some simply must have complete dryness in summer. Others are just too unruly with supplemental summer water and are better behaved with little or none.

As with any other plants, low-water plants must be grouped by water needs if in-ground irrigation is to be efficient. Even if you intend to water by hand, or not water in summer at all, when winter and spring are abnormally dry you may need to set up a sprinkler to help some plants through the summer drought. If low-water and no-water plants share the same space, one or the other will suffer when exposed to a common irrigation regime.

Climate Trends and Water Supply

Global temperatures have been rising at least since the middle of the last century, and most projections anticipate that this trend will continue. Effects of warming vary from one region of the world to another but prominently include the likelihood that precipitation will shift from snow to rain in many snow-fed watersheds. It is also likely that snow will melt faster and run off earlier, changing the timing of peak streamflow. These trends have been evident in parts of western North America during much of the past 50 years.

Snow is an important, even critical, seasonal water source for many regions with large mountain ranges, including most summer-dry climates. Snowpack accumulates in the mountains in winter when demand for water is low and releases water gradually to reservoirs as demand for water rises. Southern Europe, northwestern Africa, the eastern Mediterranean region, and western North America are all to various degrees dependent on snowpack for water supply.

Many water systems have been designed to take advantage of this precipitation pattern and are not well suited to a climate where snowfall is reduced or less reliable. Faster, earlier, and more concentrated runoff can exceed the capacity of storage, treatment, and flood control systems while also failing to recharge groundwater basins on which many regions also rely.

As with all impacts of rising temperatures worldwide, a reduction in snowpack will affect different regions and watersheds in different ways. Some regions are less dependent on snowpack than others. Some regions are warming faster than others, and rainfall trends are at best unclear. Regions that receive occasional summer rainfall or where summer-dry periods are relatively brief may fare better than those where summers are long and completely dry. Differences in population, urban development, agricultural land use, and water management also will continue to affect demand and supply.

One trend suggested by global and regional climate studies is an intensification of the natural variability of summer-dry climates. Already we are seeing more extreme and more frequent heat waves in parts of the Pacific Coast. When combined with the natural variation in precipitation, rising temperatures increase the likelihood that extremely hot summers and extremely dry winters will more often coincide, leading to intense or extended periods of drought. And if these events follow a winter of normal or high rainfall, the chances of wildfire multiply.

For the near future, the predominant effect is increased uncertainty of water supply. For the past several decades, water agencies have been addressing uncertainty by diversifying their water supply portfolios. Conservation, recycling, desalination, stormwater capture, and conjunctive use of surface and groundwater are some of the most common strategies.

Regardless of steps taken to reduce the impact, uncertainty of water supply likely will lead more often to local or regional restrictions on water use. Individuals and communities can take many steps to help protect themselves from water shortages while also contributing to the resilience of the water supply system. Graywater, cisterns, rain gardens, detention basins, and porous paving all make good use of water that otherwise may be lost to stormwater or sewer systems. All can be implemented by homeowners, homeowners' associations, or city governments, and all can have immediate and discernible effects on onsite water supplies.

Warming global temperatures may bring less snow and earlier runoff to mountains that serve as critical water supply for many summer-dry climates.

^ Flagstone pavers widely spaced maximize water infiltration.

Unwatered Spaces

Decks, patios, paths, and other hard surfaces provide usable outdoor spaces, even at the wettest times of year. Paving and paths can give year-round form and structure to the garden, providing a visual foil to formal or informal plantings. Along with unwatered and lightly watered planted areas, they also reduce water use.

Paving can be porous and open to the ground beneath wherever soils are suitable for infiltration. Properly constructed, porous paving absorbs runoff that otherwise would be sent to storm drains or to natural watercourses. Where infiltration is slow, underdrains or other means of dealing with stormwater may be required. Even porous paving should be sloped to drain, as sediments that collect between pavers can slow or prevent infiltration over time.

Brick, flagstone, or concrete pavers, cobbles, porous concrete, gravel, and decomposed granite all make good paths and patios. Recycled materials such as discarded brick and broken concrete can be attractively used and environmentally sound.

^ A gravel patio and flagstone path lead to the house and the garden.

Improving Drainage

Plants that need better drainage than your soil naturally provides can be planted on mounds or berms, in raised beds, or in soil conditioned with pumice, crushed rock, or compost. Thoroughly decomposed compost applied regularly to the top of the soil and covered with mulch can improve the structure of some soils over time. A thick mulch of gravel can provide perfect drainage at the soil surface for plants that prefer rocky soil.

If mounds or berms are constructed only of soil, they may sink to almost level in a few years. A mound built with rocks or gravel and soil will retain its shape longer, but even rocks may eventually work their way into the ground. Despite some advice you may receive, do not compact the soil beneath the mound. The key is finding the right combination of mound stability and free drainage. Boulders and smaller rocks incorporated into the design can help stabilize the mound or berm while providing plants with a cool root run.

Mounds and berms must be incorporated into an overall plan for grading and drainage, as they will alter drainage patterns by directing rainwater to other areas. Substantial changes to the landscape may require a grading permit.

Layered soils, where soils of different characteristics are found at different depths, can be particularly difficult to drain. One or more layers below the surface may be sufficiently impervious to cause water to flow horizontally rather than down. Layering can result from natural causes, but in gardens it is more often caused by grading and compaction during construction or maintenance. New homes often are left with a shallow layer of imported soil on top of mixed or inverted layers of subsoil and topsoil from the site. In such situations, the solution may involve extensive subsurface drainage or planting in raised beds.

< A rocky mound or slope is an ideal spot for plants that need fast drainage.

^ An elevated patio behind a rock wall and gravel paths below improve drainage while providing a gracious entry and a seating area in this Seattle garden.

^ Stormwater drains to a bioswale, where gravel and grasses absorb runoff and filter out pollutants from a parking lot at University of California, Davis.

Managing Stormwater

In urban and suburban areas, rainwater that runs off roofs, streets, and other impervious surfaces traditionally has been managed by centralized storm drain systems that deliver untreated water to lakes, creeks, or estuaries or by sewer systems that deliver combined water and sewage to treatment plants. As population increases and more land is covered with impervious surfaces, the volume of runoff can exceed the capacity of the stormwater system, causing sewage backups and flooding or polluting waterways with contaminants picked up from roofs, roads, and parking lots.

Many cities have addressed both pollution and flooding by installing more natural systems in public areas. Many also require onsite stormwater management for new construction wherever feasible. Some cities offer rebates or discounts to homeowners for disconnecting roof downspouts from the storm drain. Redirected runoff is instead sent to rain gardens or infiltration trenches that filter and absorb stormwater or to rain barrels or cisterns that capture and store it for later use.

Sizing and placement of any measure taken to hold and absorb stormwater on site must account for the volume of water expected from the runoff area served, the slope of the land, and the infiltration rate of the soil. Some calculations are necessary for the system to

In this San Francisco Bay Area garden, three large cisterns are filled by gravity with rainwater collected from the roof. >

function effectively and to avoid damaging structures or undermining the stability of nearby slopes. Local governments typically regulate projects larger than a specified size.

Rainwater Harvesting

Rainwater harvesting systems range from one or more rain barrels to small or large, aboveground or underground water tanks or cisterns. Cisterns can be designed to store water for both outdoor and indoor uses, but systems intended only for outdoor use are the simplest to design and install. These systems remove debris, transport water to a storage tank, and provide pressurization and filtration if required.

Cisterns are made of concrete, steel, fiberglass, or polyethylene and have a capacity of several hundred to more than 10,000 gallons. A complete system will include a tank or tanks; devices for venting, filtering, and draining; and all seals and gaskets needed for a watertight assembly. A submersible pump and pump controller will be needed if the system will be pressurized to run sprinklers.

All rainwater harvesting systems, including rain barrels, need an overflow valve to divert stormwater to the landscape or storm drain when the tank or barrel is full. Any system that relies on potable water as a backup supply also will require backflow prevention to avoid contaminating the potable supply.

Sizing a cistern requires calculation of the amount of water expected from the collection site, usually the roof, and the amount of water needed for the intended use or uses. Detailed instructions for sizing a cistern are available from local or state governments. Design by a licensed civil engineer may be required for cisterns that exceed a specified size.

^ Invasive ice plant (*Carpobrotus edulis*) can be beautiful in flower, but it smothers all other plants in its path.

WILDLAND INVADERS AND GARDEN THUGS

PLANTS THAT READILY ESTABLISH THEMSELVES in wildlands, on vacant city lots, and along rural roadsides typically are adapted to a wide range of environmental conditions. They are long lived, self-sow readily, and produce many flowers over a long season. Some bear abundant crops of fruit loved by birds and other wildlife. In other words, some of the same sorts of plants we seek out for our gardens are those most likely to take over and spread.

Most plants, however, are not invasive in wildlands or even weedy in gardens. Many plants that seed about freely in gardens do not become invasive in wildlands. Most of those that have invaded natural areas were introduced as promising ornamentals or for practical purposes such as erosion control. Few were accidental introductions.

Government agencies screen out known invaders at points of entry and attempt to identify potentially invasive plants that are already present. Most states maintain lists of their worst invasive weeds and many prohibit their sale, planting, or propagation. Local governments, nonprofit organizations, and citizen volunteers work to control invasive weeds in their communities. "Ivy pulls" and "broom bashings" are common weekend activities.

Despite such efforts, vastly expanded global trade, travel, and transport have significantly increased the threat of new and recurring plant invasions. Introductions no longer have to wait for scientific expeditions to return from faraway lands. Anyone with access to the internet can order plants or seeds from almost anywhere in the world and receive them in a matter of days.

With eradication of the worst weeds increasingly unlikely, most efforts are now turning to early detection and control. It is not always obvious which plants will become a problem. Some garden plants spread into wildlands but do not displace native plants or otherwise damage the natural ecosystem. Others spread invasively only in well-defined environments, such as beach sand along the coast. Many are invasive in some regions or parts of a region but not in others—at least not yet. There can be a long lag time between a plant's mere presence and its dominance and disruption of the ecosystem.

^ Pride of Madeira (*Echium candicans*) self-sows aggressively in the garden and in wildlands, especially along the California coast.

Experts try to predict which plants will become invasive by answering questions such as these: Is this plant invasive in other areas with a similar climate? Are related plants invasive in any similar climate? Does this plant displace native plants, harm wildlife, or increase fire frequency or intensity in areas where it has established? Does it produce large amounts of viable seed? Are seeds dispersed long distances by birds or other animals? By wind or water? By vehicles or discarded garden waste?

Because of the major role that gardens have always played in the spread of invasive plants, and because of the continued incursion of cultivated landscapes into wildlands, effective control of invasive weeds now depends heavily on the active participation of gardeners. Start by consulting lists of plants invasive in your area, learn to recognize them, and remove them if you find them on your property. Watch for plants that move from your garden to the garden next door or down the street and consider removing them from your garden as well.

Completely sterile cultivars of otherwise invasive plants may be safe, but some plants marketed as sterile may set small amounts of seed, enough to reproduce themselves. When low seed producers cross with other cultivars or with wild species, their offspring can be capable of spreading widely. If you grow these plants, watch them carefully for any tendency to spread.

^ A home in the Santa Barbara hills has a few widely spaced and well maintained trees and minimal plantings adjacent to the house to create a firewise landscape.

LIVING WITH WILDFIRE

THE GOAL OF FIREWISE LANDSCAPING IS TO REDUCE the intensity of fire and slow its advance as it nears the house. The basic principles are simple and few. Harden the target by making the house as resistant to fire as possible. Keep the area next to the house free of anything that will burn. Design and maintain planted areas farther out to provide no continuous path for fire to reach the house or move up into the tops of trees. Retain sufficient vegetation to buffer the house from airborne embers.

These principles can be implemented on the ground in many ways. Most communities are subject to state or local ordinances that specify details such as required or recommended spacing between plants and the distance from the house that regulations apply. Within these guidelines is usually a continuum of actions that meet the specified requirements.

Hardening the target may mean something as simple as screening exterior vents and closing off the space under a raised deck, or it could entail the considerable expense of replacing a shake roof and installing triple-pane windows. Clearing the area next to the house may mean removing foundation plantings and moving the woodpile uphill, or it could involve the installation of a continuous band of nonflammable materials, such as brick or flagstone patios, a concrete apron, or gravel mulch, all around the house. Interrupting pathways for the advance of fire could mean no more than pruning up trees and removing selected shrubs, or it may require a redesign of the entire landscape.

How much you do to protect your home from wildfire will depend on your assessment of risk. If you live in a community surrounded by forest or native chaparral, you likely are at some risk for wildfire. If your house is on a steep slope, near the top of a ridge, or in a canyon, your risk may be higher. If access is reasonable, you may want to focus on providing space for fire trucks and firefighters to maneuver and defend your home. If roads are narrow and winding, vegetation is dense and overgrown, water supply is less than ideal, and long driveways provide no safe access for firefighters, you may want to prepare your house to survive a wildfire on its own.

Landscaping that follows the guidelines for reducing the risk of wildfire can add usable outdoor space, simplify maintenance, and make plantings more attractive from different viewpoints. Firewise designs can help to control erosion, improve drainage, open up views, and even make the experience of approaching the house more interesting and inviting.

Dry-stacked rock walls and terraces that interrupt the path of fire also stabilize slopes, slow rainwater runoff, provide flat seating areas, and improve drainage for the many plants that prefer it. Spacing plants farther apart gives them room to show off their natural form and provides them with better air circulation. Separating planted areas with wide gravel paths simplifies maintenance while making it possible for visitors to stroll comfortably throughout the garden. Noncombustible gravel mulches conserve moisture and make it much easier to remove weeds. All of these measures also help to reduce the intensity and slow the path of approaching fire.

Experience gained from recent wildfires has added another principle: Coordinate with neighbors to extend the protected area beyond your property line. Some of the most destructive wildfires in recent years have spread explosively from one house to another, often without even reaching the tops of the tallest trees. Hardening an entire neighborhood or community may provide the best protection.

A tiled patio and sedge (*Carex pansa*) lawn can help to protect the house from airborne embers during a wildfire. >

^ A dense and diverse planting of deeply rooted trees, shrubs, and grasses, undisturbed by digging and without fertilizer or pesticides, is optimal for carbon capture.

THE CARBON CAPTURE GARDEN

CARBON CAPTURE IS WIDELY VIEWED AS A promising means of slowing global warming by reducing levels of atmospheric carbon dioxide, one of a number of gases responsible for trapping heat and warming the earth's surface. Carbon dioxide produced by industrial processes can be captured at its source and injected underground. Atmospheric carbon dioxide is naturally taken up by plants, which transform the gas into a form that can be stabilized and stored in soil.

There is much still to be learned about how carbon is stored in soil, especially long term. Also unknown is the extent to which plants, even on the grand scale of global forestry and agriculture, can help with carbon sequestration. The science to date does suggest, if only hypothetically, that both release and capture of atmospheric carbon can be influenced by the actions of farmers, foresters, ranchers, and gardeners in managing vegetated land.

The idea of the carbon capture garden is an outgrowth of regenerative agriculture, which seeks to improve productivity not through the application of ever more fertilizers but by promoting and maintaining the health of the soil. Some of the ways soil health is fostered are reduced or no tilling or digging, restrained use of fertilizers, no use of pesticides, and prevention of erosion and soil compaction. For agricultural soils that are regularly harvested and replanted, annual application of organic matter to the soil surface is also considered important.

Disturbing the soil, especially through mechanized tilling, disrupts soil aggregates, which are the basis of healthy soil structure. Aggregates can be preserved by avoiding digging and turning, especially when soil is too wet or too dry, and by mulching the soil to protect it from compaction by heavy rains. Disturbing the soil also releases carbon into the air through enhanced decomposition and directly or indirectly destroys soil microorganisms that contribute both to carbon release and to carbon storage. Pesticides kill soil organisms outright, and fertilizers, if overapplied, can reduce organism numbers and diversity.

^ The Melissa Garden, a honeybee sanctuary in Healdsburg, California, provides welcoming habitat for all organisms, including the microscopic creatures that build life-sustaining soil.

The carbon capture garden is designed and managed in ways that support soil microorganisms, maintain and improve soil structure, minimize carbon loss, and maximize carbon capture with a diverse array of deeply rooted trees, shrubs, and perennials. Perhaps not coincidentally, such a garden is also an environmentally friendly one that thrives without pesticides and fertilizers and with minimal trips to the landfill.

Combating climate warming may seem an ambitious charge to lay on the suburban front yard or shopping center parking lot, but millions of such plots of vegetated land add up to an immense acreage and a prodigious source of positive change. If the same energy and enthusiasm can be tapped for carbon capture as was directed at water conservation over the past 40 years, there is just no telling what might be accomplished.

Soil as Ecosystem

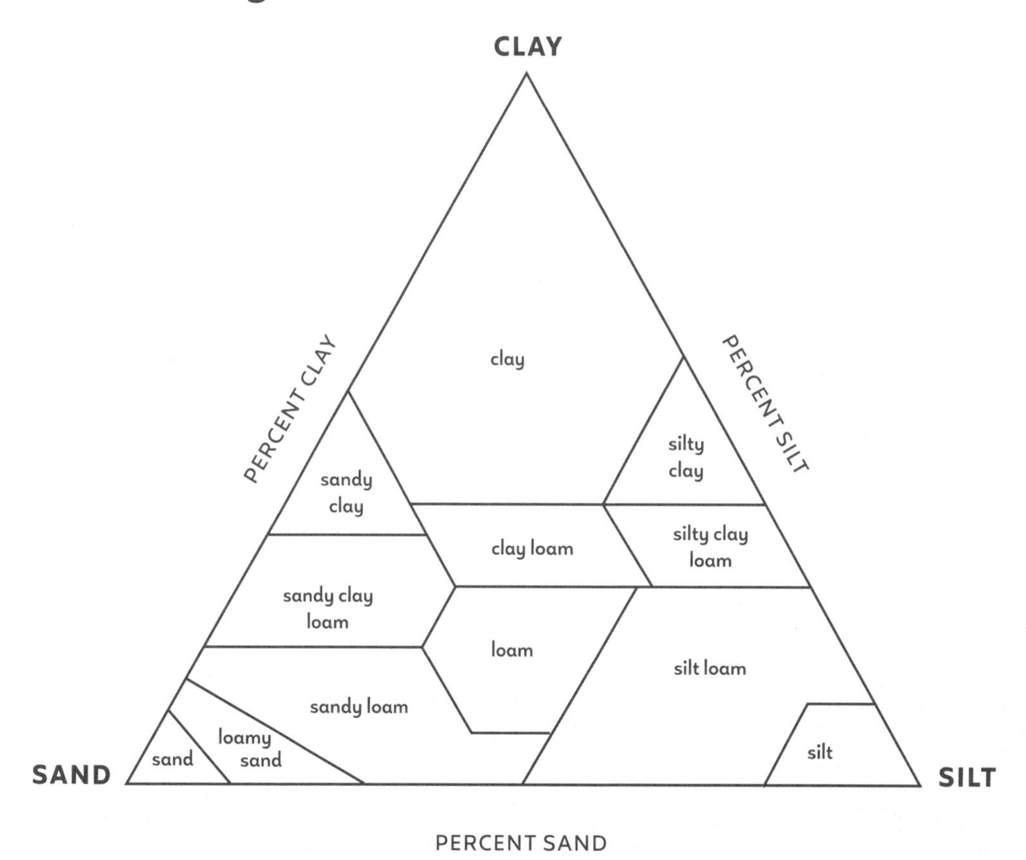

Soil is best understood as an ecosystem in which inorganic mineral solids and both living and dead organic matter interact in the presence of water and air.

The mineral component of soil consists of rock fragments, sand, silt, and clay, which are differentiated primarily by size but vary also in the minerals of which they are made. There are many different kinds of sand and many different clays. Sands, silts, and clays also vary in the ways that they influence soil functions. The amount and type of clay, for example, strongly affects the soil's ability to hold onto nutrients and water and release them slowly to plants.

The organic component of soils consists of both living and dead plant roots and residues and both living and dead soil organisms. Dead organic matter, both animal and plant, attracts and holds onto nutrients in a manner similar to clay. Organic matter also contains nutrients in its molecular structure. Living roots exude substances that feed soil organisms. Soil organisms decompose dead organic matter, in the process cycling nutrients into forms that plants can use.

Soil Texture and Structure

Most gardeners are familiar with the basics of soil texture and structure because these characteristics determine how fast or slowly a soil drains and how easy or difficult it is to insert a shovel. Soil texture and structure also determine how easily plant roots can spread into the soil, obtain water and air, and absorb nutrients.

Soil texture refers to the relative proportions of different sized mineral particles that make up the inorganic portion of the soil. Soil structure refers to the physical arrangement of soil particles and the interconnected pore spaces between them. Texture changes little over time. Structure is what gardeners can modify to create an environment beneficial for plants.

As a group, soils high in sand are coarse in texture and low in organic matter, and water and nutrients drain quickly away. Sandy soils feel gritty, even when wet. Soils high in clay are fine-textured and higher in organic matter, and they have a greater capacity to hold onto water and nutrients. Clayey soils feel sticky or slippery when wet and are hard when dry. Silty soils are intermediate in their textural characteristics.

Soil structure depends in part on the size and shape of the mineral particles—that is, soil texture—but also on the sizes and total amount of pore spaces between them. Sand particles are usually large, round, either smooth or angular, and rough surfaced, depending on how far they have traveled and how much they have been abraded. Clay particles are extremely small and layered like a stack of plates. Sand particles have relatively large pore spaces between them, while clay particles are separated by small pore spaces.

Clayey soils make up for their small pore spaces by the tendency of clay particles to cluster into aggregates. Aggregates are held together by sticky substances exuded from roots, by fine roots themselves, by fungal structures resembling roots, and by chemical compounds produced by soil organisms during decomposition of organic matter.

Healthy soils, especially healthy clayey soils, contain many different sizes of aggregates and many different sizes of pores. Tiny aggregates with small pore spaces cluster to form larger aggregates with larger pores between them. Pore spaces are habitat for soil organisms, and they are also where plant roots grow and function and where water and air both enter and leave the soil.

Living Components of Soil

Living components of soil consist of plant roots and an enormous variety of soil organisms, ranging from earthworms and insects to microscopic fungi and bacteria. Soil organisms perform many functions that assist plants, and plant roots provide nutrients to many soil organisms.

Soil organisms affect soil structure and its air- and water-holding capacity by enlarging pore spaces and binding soil particles together. They mix organic residues deeply into the soil and cycle nutrients to plants in the process of decomposition. Some organisms protect plants from pests and diseases by consuming or parasitizing harmful pathogens, while others stimulate plants to produce substances to protect themselves.

Mycorrhizal fungi aid plants directly by attaching themselves to plant roots, where they obtain sugars created by the plant by photosynthesis and provide the plant with nutrients. Plants can obtain their own nutrients from the soil, but mycorrhizal fungi are more efficient at this task.

Soil Air and Water

Living components of soil, both plant roots and soil organisms, depend on the presence of a finely tuned balance of water and air. Plant roots absorb nutrients in a water-based solution and need oxygen to function and to survive. Soil microorganisms move about in water and must have access to oxygen for respiration.

Saturated soil is a problem primarily because excess water fills soil pores and excludes air. Compaction is a problem not only because plant roots have difficulty penetrating compacted soil, but because compaction reduces or excludes both water and air. In a soil environment deficient in air and water, both plants and essential soil organisms struggle to survive.

^ The Regional Parks Botanic Garden in the hills above Berkeley is a sanctuary for plants native to California.

EMBRACING WILDNESS AND CHANGE

IT IS POSSIBLE THAT THE MOST LIFE-NEGATING ASPECT of modern landscapes is the whole idea of landscape maintenance. Landscape maintenance implies—no, insists—that landscapes must be maintained as originally designed, whatever the costs and losses.

In the service of what is called maintenance, most residential front yards and almost all commercial landscapes are bullied into submission, preserving the outlines of their original design but with virtually no sign of life. Shrubs considered too large for their space or wrongly shaped are mercilessly disfigured. Weeds and "bugs" are sprayed with pesticides. Soil, whatever its condition when first planted, is blown until nothing resembling garden soil remains. In one of the more transparent expressions of the human urge to control, all natural change in the landscape is denied.

There is just no downside to stepping back from the maintenance trap and allowing a little wildness. Stopping the pesticides brings back the pollinators as well as the microbes on which healthy soil depends. Healthy soil means that plants properly selected and placed need less water and no fertilizer to thrive. Reducing or stopping the fertilizers helps to restore populations of soil microorganisms that both feed the plants and help them to resist pests and disease.

The role of the gardener in such a setting is to foster the conditions for growth and change and to guide those changes, gently, in the continuous process of design. A garden that allows for change is never finished. If you let them, plants continue to arrange and rearrange themselves, spreading into each other and negotiating for what they need to survive. New plants appear, often installed by birds or squirrels. Sometimes the combinations of textures and colors created by the garden itself, or by its resident wildlife, are those that you may never have considered.

Relinquishing control does not mean abdication. Gardens are designed and managed in large part to serve the needs of those who make and live with them. Left to their own devices, plants may so alter the garden design that the needs of people are no longer met. The

^ A lightly tended meadow garden of perennials and grasses provides a colorful landscape in Santa Cruz.

Strawberry trees (*Arbutus* 'Marina'), New Zealand flax (*Phormium cookianum*), and native grasses intermingle in a lush landscape in Los Altos, California. >

^ A meadow garden is planted in shrubs, perennials, and grasses in San Luis Obispo.

gardener may step in at such times to restore the original design or may decide that the new arrangement is acceptable or even preferred. Gardeners change too. Needs change. Tastes change. Knowledge and experience grow. A garden that does not change with the gardener will lose the ability to delight and surprise.

Welcoming a little wildness can start small, in a portion of the garden perhaps farthest from the house, where flowers are allowed to go to seed. Here fallen leaves remain on the ground to wither and decompose, at least until spring. Some plants formerly known as weeds are encouraged to audition for a role. As you watch from a bench or hammock, it is obvious that this part of the garden is favored by birds and bees and butterflies.

We can choose to make gardens that respect the natural world, that take little from the earth that is not returned, that welcome and accommodate the presence of wildlife, and that support and nourish the life of the soil. At this critical stage of planet Earth, we are moving beyond the tenets of "live lightly" and "do no harm" to a more potent philosophy requiring positive contributions to the environment on which we depend for life itself. Will gardeners and garden designers rise to the challenge? At times it seems likely that we will.

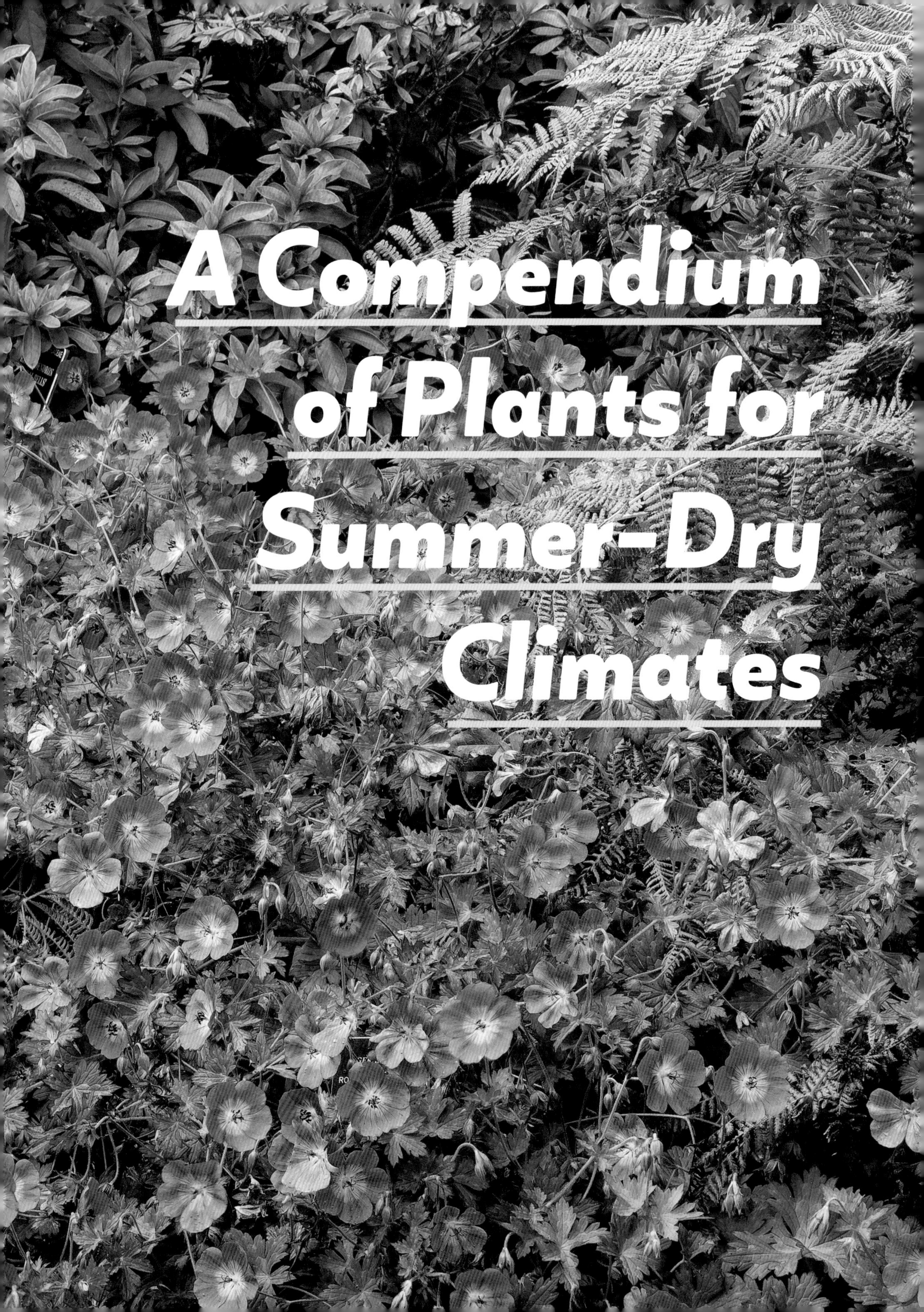
A Compendium
of Plants for
Summer-Dry
Climates

THIS COMPENDIUM provides an alphabetized catalog of photographs and brief descriptions of plants that succeed in summer-dry, winter-wet climates of the Pacific coast of North America. They should do as well in similar climates elsewhere, given the right conditions.

Plants selected for inclusion here thrive with moderate to no summer water, are attractive additions to residential or commercial landscapes, and are reasonably available in nurseries or from specialty growers, botanic gardens, or native plant sales. Hundreds of other plants could have been included, given unlimited space.

Each entry provides a brief description of the plant, information on where it grows naturally, and its needs for sun, drainage, and summer water. Codes at the end of each entry indicate U.S. Department of Agriculture (USDA) hardiness zones, Sunset climate zones, and estimated water needs from California's Water Use Classification of Landscape Species (WUCOLS) database.

Not all plants included are low water everywhere in the Pacific Coast region. Some plants that need little to no water in the Pacific Northwest or in the San Francisco Bay Area may need regular water away from the coast in southern California. If you are unsure how much water a plant will need in your garden, check with your local nursery, native plant society, or garden club.

< *Geranium* ROZANNE ('Gerwat') is one of many plants that thrive with low summer water along the Pacific coast but that may need regular water in hot-summer climates.

PLANT NAMES

Plant entries are listed alphabetically by scientific name followed by the common name or names. Each entry also lists other botanical names under which the plant may be offered in the nursery trade. The intent is to aid gardeners in locating and obtaining a particular plant, not to suggest which name is the correct one for the plant in question.

Cultivar names are provided even when a trademark is more widely used because they offer an additional means by which a cultivated variety can be identified. Cultivar names appear in single quotation marks as required by code. Trademark names appear here without quotation marks and in SMALL CAPS.

Many scientific names of plants have changed in recent years or are in the process of being changed. Botanical names are agreed upon worldwide according to rules set forth in the International Code of Nomenclature. The process of establishing new names can take many years. Many more years may pass before a new name is adopted by nurseries and by gardeners.

Occasionally the botanical name of a plant may be returned to an older name almost before the new name makes its way into the gardening lexicon. The name of the plant long known as *Diplacus aurantiacus*, or sticky monkeyflower, for example, was officially changed some years ago to *Mimulus aurantiacus* and has now been returned to *Diplacus*. Many nurseries and gardeners never made the switch.

To complicate matters further, a plant may be marketed under a trademark name that does not reveal its identity as a cultivar offered under another name. The mostly nonfruiting olive tree, *Olea europaea* ARIZONA FRUITLESS, for example, is essentially the same plant as *O. europaea* 'Swan Hill'. *Artemisia schmidtiana* SILVER MOUND is the same plant as *A. schmidtiana* 'Nana'. And as trademark names come into more common use, cultivar names are becoming increasingly enigmatic. *Vitex agnus-castus* BLUE DIDDLEY, for example, is the trademark name for a compact, blue-flowered form of the chaste tree with the cultivar name of 'SMVACBD'.

Botanical names evolve; many nurseries and even more gardeners continue to call California fuchsia, the genus *Epilobium*, by its former genus name, *Zauschneria*. >

< The botanical name for sticky monkeyflower has changed over the years from *Diplacus aurantiacus* to *Mimulus aurantiacus* and back again.

PLANT ORIGIN

It is usually helpful to know something about where and under what conditions a plant grows naturally before bringing it into the garden. Plants native to summer-dry climates are likely to be accepting of low summer water in other summer-dry parts of the world. Plants that grow naturally in sandy or rocky soils may do better in lean, well-drained soils when planted outside their native range. Plants that grow naturally in warm, dry habitats in cool, moist regions may succeed in cool, moist habitats in warmer, drier regions.

Each plant description provides as much information on plant origins as is widely known or could be hunted down. For straight species, this typically includes at least native range, or the part or parts of the world in which the species historically has been found. Environmental factors such as elevation and aspect, coastal or inland, and soil texture are provided where known.

For cultivated varieties of straight species, either wild or of garden origin, descriptions may provide information both on the native range of the species and on the general location in which the particular variant was found. For hybrids, either naturally occurring or artificially bred, information is provided on parentage or likely parentage as well as on the native range of the parent plants unless parentage is unknown. Cultivars of straight species, whether found in the wild or discovered in gardens, are described as selections to distinguish them from hybrids. This distinction is made even though, strictly speaking, all cultivars are manmade selections, as opposed to natural selections that occur in the wild.

Information on origins may not be helpful for several reasons. Some plants will thrive in conditions that are quite unlike those where they are native. Plants of species with a wide native range may come from different populations adapted to quite different environments. Then, too, gardens are not natural habitats. The conditions that the gardener can provide will not match, exactly, those in which a plant grows naturally.

Mass production of plants offered for sale through major retail chains has magnified the problem. A plant may be selected in Austin, Texas, cloned by the thousands in western Oregon, sent to southern California for growing on, and distributed to retail nurseries throughout the nation or the world. By the time it arrives at a local garden center, the native origin of the species may tell us little about the plant in hand.

Sometimes knowing where a plant was raised or how a cultivated variety was selected may be as revealing as the native origin of the species. Nurseries and breeders often grow and select plants for traits such as hardiness or heat tolerance as well as flower or leaf color. A plant native to Australia but grown in Portland and selected for its hardiness there may be more successful in Portland than a cultivar of the same plant grown and selected in southern California.

^ Jounama snow gum (*Eucalyptus pauciflora* subsp. *debeuzevillei*), from the highest mountains of southeastern Australia, is hardy in this Portland garden.

^ *Lagerstroemia* 'Tuscarora', a mildew-resistant crape myrtle with pinkish red flowers, is a good choice for gardens near the coast.

PLANT DESCRIPTIONS

Plants are described here in terms of mature size, growth habit, and leaf and flower shapes and colors. Descriptions are nontechnical, especially in describing the parts of a flower. Colorful sepals and bracts that look like petals are both called bracts. A multitude of small flowers in an elongated cluster is called a spike rather than a raceme or panicle. A flat-topped cluster of flowers is called a flat-topped cluster rather than a corymb, cyme, or umbel. Readers should consult botanical references for scientifically rigorous plant descriptions.

Mature sizes of plants, both height and width, are rough estimates only. Size varies with more than genetic makeup. Trees grown in gardens often are not as tall at maturity as they are in the wild. Shrubs may grow taller in mild-winter regions than in colder climates, and in windy locations they may hug the ground. Plants accustomed to lean soils and little rainfall may grow much larger in rich soils and with more rain or with irrigation. Mature size as well as growth rate may vary with climate and microclimate, soils and drainage, sunlight intensity and day length, and competition from other plants for water, sunlight, and nutrients.

Some plants not on any invasive species lists may nonetheless be troublesome in the garden. Watch for phrases such as "self-sows freely" or "forms thickets." Although perhaps not considered invasive in wildlands, these plants can be garden thugs that spread widely and are difficult to eradicate. If you are not willing either to control these plants or to let them have their way, it is best not to plant them.

< *Verbena bonariensis* self-sows prolifically, especially in moist soils.

WATER NEEDS

Plants are described as preferring moderate, occasional, infrequent, or no summer water. Moderate is roughly defined here as once a week, occasional as every two weeks, and infrequent as once a month.

The terms are relative. Plants are described as needing moderate water if they do best with more water than another plant that needs occasional water in the same location. If a plant is said to prefer occasional to infrequent or no summer water, this means that occasional water may be needed in warmer, drier locations but infrequent to no summer water may be needed where summers are cooler and humidity is high. A plant needing no summer water in Vancouver may need deep watering twice a month in Portland, weekly watering in Sacramento, and somewhere between monthly and no summer water in San Francisco.

A plant also may have different needs for water in two different gardens in the same city, or in two different spots in the same garden. Water needs will be lower in afternoon shade than in all-day full sun. If that shade is provided by a mature tree, competition from tree roots may increase the water needs of plants beneath it. Other site conditions to consider are the heat reflected by a hot stucco wall, the drying effect of a persistent breeze, or the moisture retained by several inches of mulch.

Estimates of water needs assume that the plant is well established, which can take several years. Perennials may need more water only for the first summer and accept a lesser amount in the second year. Trees may need moderate water the first summer, occasional water the second, and infrequent water the third or fourth year before settling down to a regime of no summer water in year five or six.

Codes used in California's Water Use Classification of Landscape Species (WUCOLS) database are included for each catalog entry as another tool for comparing water needs. WUCOLS provides estimates of the water needs of several thousand plants in six different climates of California. The WUCOLS codes are: VL (very low), L (low), M (moderate), H (high).

WUCOLS codes are useful well beyond California, as they suggest the relative needs of plants for water in different climates. If you know from experience how much water one plant needs in your garden, other plants with the same WUCOLS code may thrive with similar amounts. The codes were assigned based on the amount of water needed to maintain an established plant's health, appearance, and growth rather than how much is needed for the plant to survive. A plant may succeed with less water than its code suggests but generally will not need more.

For our purposes, a WUCOLS code of L/M means that the plant thrives with occasional water in cooler northern or coastal climates but needs moderate summer water in hotter southern or inland climates. A code of L/M/VL means that WUCOLS has assigned the plant to three different water use categories for different regions. A single code of L means that the plant has low water needs in all regions for which it was evaluated.

DRAINAGE

Each entry includes a rough estimate of the plant's drainage needs or preferences. Excellent drainage means that a hole filled with water drains completely in a few minutes. If the hole drains within an hour or two, the soil has reasonably good drainage. Poorly drained soil may take several to many hours to drain.

Most plants need at least reasonably well-drained soil for optimum growth. Drainage ensures that the soil is properly aerated. It also affects how much water is available for use by plants. Maintaining the right amounts of air and water in the soil is a fundamental determinant of plant health and survival.

SUN OR SHADE

Preferences for sun or shade suggest the relative amount of sun a plant needs or can tolerate. This will vary with elevation, latitude, time of year, and time of day as well as the genetics of the plant.

Full sun means direct sunlight for much of the day, but the number of hours a plant can accept will vary with the intensity of the sun. Sunlight is most intense at higher elevations, in more southerly latitudes, on south-facing slopes, in midsummer, and in the middle of the day. Plants that thrive in full sun in northern climates or along the coast may need part shade or afternoon shade in hotter southern or interior locations.

Part shade means filtered sunlight, such as under the partly open canopy of a tree. Afternoon shade provides protection from direct sun during the hottest part of the day. Many plants prefer some shade in the afternoon in hot-summer climates, even if their nursery labels advise full sun. Plants described as needing full sun also may need temporary shading in summer while becoming established.

Agave attenuata (here 'Boutin Blue') is particularly sensitive to intense afternoon sun; many plants do better with some afternoon shade, even if their nursery labels advise full sun.

CLIMATE AND MICROCLIMATE

Each entry ends with codes for USDA hardiness zones and Sunset climate zones whenever that information is available. Each is useful for different reasons and neither is sufficient to determine with certainty whether a plant will succeed in your garden. It is always best to check with a local nursery or garden club to determine whether a particular plant or cultivar is suited to your climate and microclimate.

USDA hardiness zones are based on average annual minimum winter temperatures over a 30-year period, adjusted to some extent for elevation and position with respect to large bodies of water. The 13 zones are separated by an increase in temperature of 10 degrees F., with lower numbers being colder. Each zone is further divided into subzones a or b, representing 5-degree increments. USDA codes are commonly provided on plant labels, in nursery catalogs, and in online marketing because they are recognized throughout North America and in many other parts of the world.

Sunset climate zones take into account not only winter low temperatures but summer highs, wind, humidity, timing and amount of rainfall, proximity to the ocean, and length of the growing season. Of the 50 Sunset zones in the United States, 24 apply to the western states. These codes are more precise descriptors of climate than the USDA zones and they reveal more about a plant's adaptability. They are not as widely used by growers and nurseries, especially outside the western states, and often do not appear on plant labels or in nursery catalogs.

Both the USDA and Sunset codes are guidelines only. Many plants have been thoroughly field tested to determine their zones, but many others have been assigned to zones based on the best estimates of those who have grown them. USDA rankings in particular vary, sometimes wildly, from one grower, nursery catalog, or publication to another. The USDA describes climate zones but, unlike Sunset, does not assign plants to the zones the agency has established, leaving that task to others.

It is important to remember that both Sunset and USDA climate zones are averages and colder temperatures do occur. Plants rated zone 8 may freeze in zone 8 in years when low temperatures fall below average, especially if nighttime low temperatures fail to rise the following day. Alternatively, plants rated zone 8 may do fine in zone 7 for many years in a row, only to be cut down by an unusually hard freeze.

If you want to experiment with plants rated outside your zone, it is wise to choose trees and long-lived shrubs conservatively and save the experiments for short-timers that are more easily replaced.

^ Phormiums can grow larger than expected where conditions are favorable.

Abutilon palmeri

PALMER'S INDIAN MALLOW

Evergreen shrub, 3–5 feet tall and wide, with velvety, heart-shaped, gray-green leaves and golden yellow to orange, bell-shaped flowers in spring and summer. Native to dry, rocky soils, usually on east-facing slopes below 2,400 feet, in mountains of southern California, Arizona, and northwestern Mexico. Sun to part shade, well-drained soils, infrequent to no summer water. Needs summer heat. USDA: 9–11 Sunset: 8–9, 11–13 WUCOLS: L

^ *Abutilon palmeri*

Acacia

ACACIA

Evergreen shrubs and small trees, fast growing, with fine-textured, narrow, green or gray-green leaves and often fragrant, yellow flowers in midwinter or early spring, followed by beanlike seedpods. Some acacias are on invasive species lists, including *A. dealbata*, *A. longifolia* (*A. latifolia*), *A. melanoxylon*, and *A. baileyana*, and all should be watched for tendencies to spread. Sun to part shade, most well-drained soils, occasional to infrequent or no summer water. May be short lived.

A. boormanii, Snowy River wattle, 10–15 feet tall and 8–12 feet wide, with silvery gray bark and gray-green leaves. Suckers and may form thickets. Native to the mountains of southeastern Australia. Occasional summer water. USDA: 8–12 Sunset: 8–9, 12–24 WUCOLS: L

^ *Acacia boormanii* in two seasons

A. cognata, river wattle, 15–30 feet tall and wide, with narrow, drooping, bright green leaves on pendulous branches. Native to coastal southeastern Australia. May be offered as *A. subporosa* var. *linearis*. COUSIN ITT ('ACCOG01') is 2–3 feet tall and 4–6 feet wide with the same pendulous habit; may be offered as 'Mini Cog'. Best in southern California or along the coast with occasional summer water. Part shade inland. USDA: 9–11 Sunset: 16–24 WUCOLS: L/M

Acanthus mollis 

^ *Acacia cognata* COUSIN ITT

A. covenyi, blue bush, 12–20 feet tall and 10–15 feet wide, with silvery blue, narrowly elliptical leaves on dark gray stems. Native to coastal mountains of southeastern New South Wales, Australia. Infrequent summer water. USDA: 9–10 Sunset: 8–9, 12–24 WUCOLS: L

A. redolens, prostrate acacia, 1–4 feet tall and 10–15 feet wide, with gray-green, lance-shaped leaves. Native to coastal and inland areas of southern Western Australia. DESERT CARPET and 'Low Boy' are 1–2 feet tall. Needs ample room to spread. Infrequent to no summer water along the coast, moderate to occasional water inland. Weedy and potentially invasive in parts of southern California. USDA: 9–11 Sunset: 8–9, 12–24 WUCOLS: L/VL

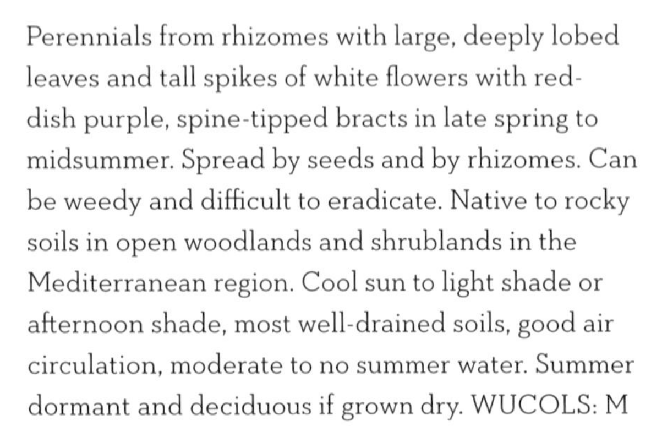

Acanthus

BEAR'S BREECH

Perennials from rhizomes with large, deeply lobed leaves and tall spikes of white flowers with reddish purple, spine-tipped bracts in late spring to midsummer. Spread by seeds and by rhizomes. Can be weedy and difficult to eradicate. Native to rocky soils in open woodlands and shrublands in the Mediterranean region. Cool sun to light shade or afternoon shade, most well-drained soils, good air circulation, moderate to no summer water. Summer dormant and deciduous if grown dry. WUCOLS: M

A. mollis, 3–4 feet tall and 2–3 feet wide, with glossy, dark green leaves. Native from Portugal to Italy and northwestern Africa. 'Hollard's Gold' has chartreuse leaves. USDA: 7–10 Sunset: 5–24

A. spinosus, 2–4 feet tall and wide, with finely cut, dark green or silvery green leaves tipped with prickly spines. Native from Italy to western Turkey. 'Whitewater' has variegated leaves of green and creamy white. USDA: 6–10 Sunset: 4–24

Acca sellowiana
PINEAPPLE GUAVA

Evergreen shrub or small tree, slow growing to 8–20 feet tall and wide, with dark green, oval leaves, silvery beneath, and showy white and pink flowers with brilliant red stamens in spring or early summer. Flowers are followed by rounded to pear-shaped, gray-green fruit. Fruit and leaf drop can be messy. Native to the highlands of southern Brazil to Paraguay, northern Argentina, and Uruguay. Sun to light shade, most well-drained soils, occasional to infrequent summer water. Moderate water if grown for the purplish green fruits. May be offered as *Feijoa*. USDA: 8–10 Sunset: 7–9, 12–24 WUCOLS: L/M

^ *Acca sellowiana*

Achillea
YARROW

Perennials, mat forming to upright and clump forming, with soft, finely divided, usually aromatic, green to gray-green leaves and broad, flat-topped clusters of tiny summer flowers on leafy stems. Can be weedy. Spread by rhizomes and by seed. Sun to part shade, most well-drained soils, moderate to occasional or infrequent summer water. USDA: 4–8 Sunset: 1–24, A1–3

^ *Achillea* 'Moonshine'

A. filipendulina, fernleaf yarrow, low basal clump of bright green leaves, 2–3 feet wide, and golden yellow flowers on 3- to 4-foot stems. Native to gravelly or sandy soils, from Iran and Afghanistan north to the Caucasus and east to Central Asia. 'Gold Plate' has gray-green leaves and especially broad flower clusters. WUCOLS: L

A. millefolium, common yarrow, variable, from upright to mat-forming and 2–3 feet wide, with medium green to gray-green leaves and white to pinkish white flowers on 2- to 3-foot stems. Spreads, sometimes aggressively. Native to a wide range of habitats in temperate regions of Asia, Europe, and much of North America. 'Island Pink', from the Channel Islands in southern California, has green leaves and rose-pink flowers. 'Calistoga', from the Coast Ranges in northern California, has silvery gray leaves and creamy white flowers. WUCOLS: L/M

A. 'Moonshine', 1–2 feet tall and wide, with silvery gray-green leaves and lemon-yellow flowers on 2- to 3-foot stems. Believed to be a hybrid between *A. clypeolata*, a silvery gray-leaved species native to the Balkan Peninsula, and *A. aegyptiaca* (*A. taygetea*), a low-growing, gray-leaved species from Greece. WUCOLS: N/A

A. tomentosa, woolly yarrow, dense mat, 3–6 inches tall and 12–18 inches wide, with woolly, green or gray-green leaves and small clusters of yellow flowers on short stems. Native to southern Europe and western Asia. 'Maynard's Gold' has green leaves and bright yellow flowers. 'King George' has gray-green leaves and creamy yellow flowers. WUCOLS: L/M

Adenanthos
WOOLLYBUSH

Evergreen shrubs, prostrate to erect and treelike, with silvery gray-green leaves and small, red or pinkish red, tubular flowers at almost any time of year. Sun to part shade, good to excellent drainage, occasional summer water. Best near the coast but fine inland with afternoon shade. Sunset: 8–9, 14–24 WUCOLS: L

A. cuneatus, coastal jugflower, prostrate or upright to 3–6 feet tall and wide, with scalloped, wedge-shaped leaves, new growth pinkish red. Native to sandy soils along the south coast of Western Australia. 'Coral Drift', 2–3 feet tall, has bright coral-pink new growth. 'Coral Carpet' is similar but 6–10 inches tall. 'Flat Out' is even lower and spreads about 3 feet wide. USDA: 8–10

A. sericeus, coastal woollybush, upright shrub or small tree, 8–12 feet tall and wide, with fine-textured, needlelike but soft and silky leaves, new growth bright green. Native along the south coast of Western Australia. USDA: 9–11

^ *Adenanthos sericeus*

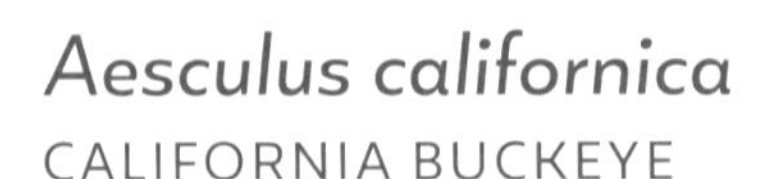

Aesculus californica
CALIFORNIA BUCKEYE

Deciduous tree or large shrub, 15–40 feet tall and wide, usually multitrunk, with silvery gray bark, medium green leaves, bright green when new, upturned clusters of fragrant, creamy white flowers at branch ends in spring, and huge, shiny, brown seeds from pear-shaped pods in fall. Leaves tend to turn yellow and drop in midsummer. Native to woodlands below 4,000 feet in the Siskiyou Mountains, Coast Ranges, and Sierra Nevada foothills from southwestern Oregon to northern Los Angeles County. Best form is in sun, but grows well in filtered shade, where it needs no summer water. Occasional water may postpone leaf drop. USDA: 7–10 Sunset: 3–10, 14–24 WUCOLS: VL/L

^ *Aesculus californica*

^ *Agastache rupestris*

Agastache

HUMMINGBIRD MINT

Perennials with aromatic, green to gray-green, narrowly linear, lance-shaped, or oval leaves and upright spikes of long-lasting, tubular, often multicolored flowers in summer. Most self-sow. Plants listed here are native to the southwestern United States and northwestern Mexico and thrive with moderate water and excellent drainage in sun or light shade. Other species and many hybrids may need regular water.

A. aurantiaca, orange hummingbird mint, 2–3 feet tall and 2 feet wide, with gray-green leaves and orange flowers with purple sepals. Native to mountains of northern Mexico. CORONADO ('P012S'), with yellow flowers suffused with orange, and 'Apricot Sprite', with bright orange flowers, are more compact. USDA: 6–9 Sunset: 3–24 WUCOLS: M

A. cana, Texas hummingbird mint, 2–3 feet tall and 1–2 feet wide, with gray-green leaves and pink flowers with purple sepals. Native to dry slopes and canyons in mountains of western Texas and southern New Mexico. 'Rosita', with rose-pink flowers, and SONORAN SUNSET ('Sinning'), with purple flowers, are more compact. USDA: 5–9 Sunset: 2–24 WUCOLS: L/M

A. rupestris, sunset hyssop or licorice mint, 2–3 feet tall and wide, with narrowly linear, silvery gray-green leaves and pinkish orange flowers with lavender sepals. Native to mid-elevations in the mountains of Arizona, New Mexico, and northern Mexico. 'Apache Sunset' is shorter. USDA: 5–9 Sunset: 1–24 WUCOLS: L/M

Agave
AGAVE

Succulent or semi-succulent rosettes of green to blue-green or blue-gray leaves, usually with sharp marginal teeth and a terminal spine. Most flower only once, at maturity, after which the rosette dies. Many produce offsets or young plants that can be pried away and replanted or left to form a slowly expanding colony. Most agaves need full sun along the coast to part shade or afternoon shade inland, excellent drainage, and occasional to infrequent summer water. Most are at least somewhat frost tender; some are quite cold hardy where protected from winter rains. Plant tilted to let water drain rapidly away. WUCOLS: L/VL

A. attenuata, foxtail agave, rosette 2–3 feet tall and wide, with flexible, pale green leaves lacking teeth or spines, eventually forming a 4- to 5-foot stem. Chartreuse flowers on an arching 10- to 12-foot stalk. Offsets freely. Native to rocky outcrops between 6,000 and 8,000 feet on the west-facing side of the mountains of central Mexico. 'Boutin Blue' and 'Nova' have blue-gray leaves. Frost tender and leaves burn in hot sun. USDA: 9–11 Sunset: 13, 20–24

^ *Agave attenuata*

A. 'Blue Glow', 1–2 feet tall and 2–3 feet wide, almost perfectly symmetrical, with stiffly upright, chalky blue-green leaves with finely toothed margins and a terminal spine. Pale yellow flowers after 8–10 years. Does not form offsets. Hybrid of *A. attenuata* and *A. ocahui*, a plant from mid-elevation desert regions of northeastern Mexico. USDA: 8–11 Sunset: 9, 13–24

^ *Agave* 'Blue Glow'

A. bracteosa, candelabrum or spider agave, slow-growing rosette to 1–2 feet tall and wide, with pale green to gray-green, flexible and arching, spineless leaves. Creamy white flowers on a 3- to 5-foot stalk. Offsets are produced and flowering rosette may persist for some time. Native to steep, rocky cliffs between 3,000 and 5,500 feet in the mountains of northeastern Mexico. Needs occasional summer water inland. USDA: 8–10 Sunset: 7–9, 12–24

A. filifera, threadleaf agave, rosette to 1 foot tall and 2 feet wide, with stiff, narrow, dark green, upturned leaves edged in white with curly, white filaments along toothless margins and a sharp terminal spine. Greenish purple or purple-tinted yellow flowers on a dark purple, 6- to 10-foot stalk. Usually offsets, forming a small colony. Native to rocky, forested slopes between 3,000 and 8,000 feet in mountains of central Mexico. May be offered as *A. filamentosa*. USDA: 9–11 Sunset: 12–24

A. havardiana, Big Bend agave, rosette 2–3 feet tall and 3–4 feet wide, with short, broad, silvery blue-gray or gray-green leaves with marginal teeth and a sharp terminal spine. Greenish yellow flowers on a 15-foot stalk. Few offsets. Native to grasslands and rocky woodland openings between 4,000 and 6,500 feet in western Texas, New Mexico, and northern Mexico. USDA: 7–10 Sunset: 7–24

A. parryi, Parry's agave or mescal, rosette 2–3 feet tall and 3–4 feet wide, with short, broad, light gray-green to blue-gray leaves with reddish brown marginal teeth and a sharp terminal spine. Bright yellow flowers on a branched 15-foot stalk. May form offsets. Native to grasslands, chaparral, and woodlands between 4,000 and 8,000 feet in Arizona, New Mexico, and northern Mexico. 'Flagstaff', with steel blue leaves, is a high-elevation selection. USDA: 7–10 Sunset: 2b, 3, 6–24

A. parryi subsp. neomexicana is a smaller plant with narrower leaves that offsets freely. Native to rocky slopes and grasslands between 1,400 and 7,000 feet in western Texas and southeastern New Mexico. One of the most cold-hardy agaves where perfect drainage can be provided. USDA: 6–10 Sunset: N/A

A. schidigera, threadleaf agave, rosette 18 inches tall and 2 feet wide, with dark green leaves with a terminal spine and curly, white filaments on the margins. Native to the mountains of north-central Mexico. Does not form offsets. May be offered as a subspecies of *A. filifera*. USDA: 8–11 Sunset: 9, 13–24

< *Agave schidigera*

^ *Agonis flexuosa*

Agonis flexuosa

PEPPERMINT WILLOW

Evergreen tree, fast growing to 25–40 feet tall and 15–25 feet wide, with a weeping habit, deeply fissured brown bark, narrowly linear, gray-green to olive-green aromatic leaves, and clusters of small, white flowers in spring or summer. Leaf drop can be messy. Native to sandy soils, a few miles inland from the coast, in southwestern Western Australia. 'Jervis Bay Afterdark' ('After Dark') is smaller, with dark burgundy, almost black, leaves. Sun to part shade, most well-drained soils, moderate to occasional summer water. Best near the coast. USDA: 9b–11 Sunset: 15–17, 20–24 WUCOLS: L

Allium

WILD ONION

Perennials from bulbs, with one to several narrow, grasslike leaves, and clusters of tiny, star-shaped or cup-shaped flowers on leafless stalks. Strong onion odor from all parts of the plant. Leaves of most die back with peak bloom. May self-sow. Some ornamental onions flower in summer and need water while blooming. Those listed here flower in spring, are dormant or partially dormant in summer, and need no summer water. Sun to part shade, good drainage. WUCOLS: L/VL

A. amplectens, narrow-leaved onion, with small, pale pink or white flowers in flat-topped clusters. Native to dry, rocky slopes and meadows from southwestern British Columbia to central California. USDA: 6b–9 Sunset: N/A

A. cristophii, star of Persia, with large, spherical clusters of lavender to purple flowers. Strap-shaped, gray-green, basal leaves persist during flowering. Native to Iran, Turkey, and Turkmenistan. May be offered as *A. albopilosum*. USDA: 5–8 Sunset: 1–24

A. schubertii, tumbleweed onion, with strap-shaped, bluish green, basal leaves and huge, spherical clusters of rose-purple flowers that bloom at the ends of individual stalks of varying lengths. Native to the eastern Mediterranean region from Turkey to Lebanon, Israel, and Libya. USDA: 5–9 Sunset: N/A

A. unifolium, one-leaf allium, with several flattened green leaves and roundish clusters of pink to pinkish lavender flowers. Native to grasslands in coastal mountains of Oregon, California, and Baja California. Can be weedy. 'Wayne Roderick' has particularly deep lavender flowers. USDA: 5–8 Sunset: 3–9, 14–24

^ *Allium cristophii* seedheads

^ *Allium unifolium*

Aloe

ALOE

Succulent rosettes of broadly or narrowly lance-shaped, sword-shaped, or roughly triangular leaves, often with marginal teeth, and tubular or bell-shaped flowers on leafless stalks. Most are frost tender; some are fairly hardy if winter rains drain rapidly away. Hundreds of named cultivars, many of unknown or complicated parentage. Sun to light shade or afternoon shade, good to excellent drainage, occasional to infrequent summer water. USDA: 9–11 WUCOLS: L

A. arborescens, torch aloe, multiple rosettes of narrow, softly toothed, gray-green or bluish green leaves on a branching stem, eventually forming a shrubby plant 6–8 feet tall and wide. Red-orange flowers on 2-foot, unbranched stalks in fall and winter. Native to rocky soils from sea level to 8,000 feet in southern Africa from east of Cape Town to summer-rainfall Mozambique. 'Lutea' and 'Yellow Torch' have yellow flowers. Sunset: 8–9, 13–24

A. 'Blue Elf', rosette 1 foot tall and spreading 2 feet wide, with narrow, erect, silvery blue-gray leaves and orange flowers in winter. Hybrid of uncertain parentage, possibly involving *A. humilis*. May be offered as 'Blue Boy' or as California aloe. Good choice for hot sites. Accepts shade. Sunset: 8–9, 12–24

A. brevifolia, short-leaved aloe, small, stemless rosette 3–4 inches tall and wide, with short, gray-green leaves with white marginal spines. Orange flowers on 2-foot, unbranched stalks in late spring. Slowly forms a clump to 1 foot tall and 2 feet wide. Native to dry, rocky slopes from near sea level to 500 feet along the coast in Western Cape Province, South Africa. Good choice for smaller gardens. Sunset: 8–9, 12–24

A. marlothii, mountain aloe, rosette 4–5 feet tall and wide, with gray-green leaves covered with short, red spines, slowly forming an unbranched stem to 6 feet tall. Yellow, orange, or red flowers on a horizontally branching stalk in late fall to winter. Native to summer-rainfall regions of southern Africa from northeastern South Africa to Mozambique. Sunset: N/A

A. striata, coral aloe, rosette 18 inches tall and 2 feet wide, with broad, pale green to gray-green, subtly striped leaves with pale red, toothless margins. Coral-red flowers on a 2-foot, branching stalk in winter or early spring. May produce offsets, forming a small cluster. Native to dry, rocky slopes from 800 to 7,300 feet in Eastern and Western Cape provinces, South Africa. Sunset: 8–9, 12–24

^ *Aloe arborescens*

^ *Aloe* 'Blue Elf'

Aloysia citrodora

LEMON VERBENA

Evergreen shrub, upright and fast growing to 6–8 feet tall and 4–6 feet wide, with narrowly lance-shaped, bright green, aromatic leaves and loose clusters of tiny white or pale lilac flowers in summer and fall. Native to much of South America from Peru to Brazil and from Chile to Uruguay. Deciduous in cold-winter climates. Full sun, most soils, occasional summer water. Best along the coast. May be offered as *A. citriodora*, *A. triphylla*, or *Verbena triphylla*. USDA: 8b–10 Sunset: 9–10, 12–21 WUCOLS: L

^ *Aloysia citrodora*

Alyogyne huegelii

BLUE HIBISCUS

Evergreen shrub, upright to 6–8 feet tall and wide, with deeply lobed, rough-textured, dark green leaves and large, pale blue to bluish lilac or purple flowers almost year-round in mild-winter climates. Native to sandy or gravelly soils near the coast in Western Australia. 'White Swan' has purple-tinged, white flowers. Sun to part shade, fast drainage, occasional to infrequent or no summer water. Tip prune any time for a more compact plant. USDA: 9b–11 Sunset: 13–17, 20–24 WUCOLS: L

^ *Alyogyne huegelii*

Amaryllis belladonna

NAKED LADY

Perennial from bulb, 1 foot tall and 2 feet wide, with medium green, strap-shaped leaves that emerge in late fall or winter and die back in spring before summer dormancy. Large, pale pink, trumpet-shaped flowers on 2- to 3-foot, leafless stems in late summer or fall. Native to Western Cape Province, South Africa. Sun to part shade or afternoon shade, most well-drained soils, no summer water. Hybrids with *Brunsvigia*, a bulbous plant in the same family, extend the color range of flowers to pure white and red. Known as ×*Amarygia*, these are equally easy and hardy. USDA: 8–10 Sunset: 4–24 WUCOLS: VL

^ *Amaryllis belladonna* (center)

^ *Andropogon gerardii*

^ *Anemanthele lessoniana*, spring foliage

^ *Anemanthele lessoniana*, fall foliage

Andropogon gerardii

BIG BLUESTEM

Warm-season bunchgrass, narrowly upright to 3–5 feet tall and 2–3 feet wide, with bluish green leaves that turn coppery red in fall and purplish red, late-summer flowers on 5- to 7-foot stems. Spreads by rhizomes and may form a rough sod in moist locations. Native to grasslands and open woodlands in much of midwestern and eastern North America. WINDWALKER ('PWIN01S'), a selection from southern Colorado, has blue-gray leaves that turn burgundy in fall. Full sun to part shade, most soils, moderate to occasional summer water. Cut back in late winter to renew. USDA: 4–9 Sunset: 1–9, 14–24 WUCOLS: L

Anemanthele lessoniana

NEW ZEALAND WIND GRASS

Cool-season bunchgrass, 2–3 feet tall and wide, with arching, fine-textured, dark olive-green leaves that take on coppery orange tones in fall. Airy, purple-tinged summer flowers. Native to New Zealand, usually along streams or in partly shaded forest openings from sea level to 1,500 feet. Cool sun to light shade, most well-drained soils, moderate to occasional summer water. Best near the coast, where it colors up well in full sun. May be offered as *Agrostis* or as *Stipa arundinacea*. USDA: 8–10 Sunset: 14–24 WUCOLS: L/M

^ *Anigozanthos* 'Bush Ranger'

^ *Arbutus* 'Marina'

Anigozanthos
KANGAROO PAWS

Perennials from rhizomes, 1–3 feet tall and wide, with strap-shaped, green or gray-green leaves and softly hairy, upright, tubular summer flowers on leafless stems, usually well above the leaves. Native to dry, sandy soils in southwestern Western Australia. Some species are dormant in summer and some cultivars need regular water. All need protection from heavy winter rains. 'Bush Ranger', a hybrid involving *A. flavidus* and *A. humilis*, is 12–18 inches tall and wide, with red flowers on branched 2-foot stalks. Sun to part shade, excellent drainage, good air circulation, moderate to occasional summer water. USDA: 9–11 Sunset: 15–24 WUCOLS: L/M

Arbutus
MADRONE

Evergreen trees and large shrubs with leathery, dark green leaves, colorful peeling bark, clusters of white to pale pink, urn-shaped flowers in spring, and red or orange berries in fall. Year-round shedding of bark, leaves, flowers, and fruit can be messy. Sun to light shade or afternoon shade, fast drainage, good air circulation, occasional to infrequent or no summer water.

A. 'Marina', 25–40 feet tall and 25–30 feet wide, with large, dark green leaves, bronzy when new, and large clusters of rose pink flowers. Hybrid of unknown parentage, believed to be either a selection of *A.* ×*andrachnoides*, a naturally occurring hybrid of *A. unedo* and the eastern Mediterranean *A. andrachne*, or a hybrid of *A.* ×*andrachnoides* and *A. canariensis*, native to the Canary Islands. May be offered as *A.* ×*reyorum*, but usually offered simply as *A.* 'Marina'. Best in full sun. USDA: 9–10 Sunset: 8–9, 14–24 WUCOLS: L/M

A. menziesii, Pacific madrone, upright or spreading, often multitrunk, slow growing to 20–80 feet tall and 15–40 feet wide. Native from sea level to 5,000 feet from coastal southwestern British Columbia to the western slopes of the northern Sierra Nevada and the coastal mountains of northern San Diego County. Not easy in cultivation. Needs perfect drainage. Best near the coast. USDA: 7–10 Sunset: 4–7, 14–19 WUCOLS: L

A. unedo, strawberry tree, rounded and usually multitrunk, 10–20 feet tall and wide. Red berries from the previous year persist on the tree along with new orange berries and pinkish white flowers. Native to the Mediterranean region from Portugal east to Turkey and Israel and from coastal northern Morocco to Tunisia. 'Compacta' is 6–8 feet tall and wide. USDA: 8–10 Sunset: 4–24 WUCOLS: L

Arbutus unedo

Arctostaphylos
MANZANITA

Evergreen shrubs, prostrate to midsized or tall, most with colorful, peeling bark and a picturesque, contorted growth habit. Leaves are leathery and green to gray-green or bluish green. Pendant clusters of small, white or pink, urn-shaped flowers in winter or early spring are followed by small, round, berrylike fruit. Sun to part shade, good to excellent drainage, good air circulation, occasional to infrequent or no summer water.

A. 'Austin Griffiths', upright to 8–12 feet tall and 6–8 feet wide, with medium green leaves and pink-tinged white flowers. Hybrid of garden origin from coastal central California believed to be a cross between *A. densiflora* 'Sentinel' and *A.* 'Dr. Hurd'. Best with some afternoon shade. USDA: 7–10 Sunset: 4–9, 14–24 WUCOLS: L

A. bakeri, Baker's manzanita, upright to 6–10 feet tall and 6–8 feet wide with dark gray-green leaves and pink flowers. Native to chaparral and woodland openings in the North Coast Ranges in Sonoma County, California. Usually available as 'Louis Edmunds'. Best near the coast. Afternoon shade inland. USDA: 7b–10 Sunset: 4–9, 14–17 WUCOLS: L

A. canescens, hoary manzanita, 3–6 feet tall and wide, with oval to rounded, silvery blue-green leaves and pink-tinged, white flowers. Native to dry rocky ridges, brushy slopes, and open forests in the coastal mountains of northern California and southern Oregon. Subspecies *sonomensis*, Sonoma manzanita, may be offered as *A. sonomensis*. USDA: 6b–10 Sunset: N/A WUCOLS: L/VL

^ *Arctostaphylos densiflora*

A. densiflora, Vine Hill manzanita, 3–4 feet tall and 5–7 feet wide, with dark green leaves and white or pale pink flowers. Native to sandy soils in chaparral in Sonoma County, California. Several natural hybrids involving *A. densiflora* are widely grown. 'Howard McMinn', 6–10 feet tall and wider than tall, is exceptionally adaptable. 'Sentinel' is similar but more upright. USDA: 8–10 Sunset: 7–9, 14–21 WUCOLS: L

A. 'Dr. Hurd', to 10–15 feet tall and wide, with large, light green leaves and white flowers. Hybrid of garden origin discovered in Portola Valley, California, likely involving *A. manzanita* and *A. stanfordiana*. Adaptable. May be offered as a selection of *A. manzanita*. USDA: 8–10 Sunset: 4–9, 14–24 WUCOLS: L/VL

A. edmundsii, Little Sur manzanita, less than 1 foot to 4 feet tall and 4–10 feet wide, with glossy, dark green leaves, new growth edged with red, and white to pale pink flowers. Native to coastal bluffs, away from salt spray, in Monterey County, California. 'Carmel Sur' is 1 foot tall and 6–8 feet wide, with gray-green leaves, bright green when new, and few if any flowers. Best along the coast or with afternoon shade inland. USDA: 8–10 Sunset: 6–9, 14–24 WUCOLS: L/M

A. glauca, bigberry manzanita, 6–15 feet tall and wide, with pale gray-green or blue-gray leaves and white or pink flowers. Native to chaparral and woodland openings below 4,500 feet, both along the coast and inland, from central California south into Baja California. Best with some shade in the hottest locations. Good choice for southern California. USDA: 7–10 Sunset: 4–9, 14–24 WUCOLS: VL

A. hookeri, Monterey manzanita, variable, 1–4 feet tall and 3–12 feet wide, with small, glossy, bright green leaves and white or pale pink flowers. Native to coastal bluffs and open woodlands below 1,000 feet along the central and northern California coast. 'Monterey Carpet', 1–3 feet tall and 6–8 feet wide, has gray-green leaves with serrated margins and white flowers. 'Wayside' is a little taller, with bright green leaves and white flowers. Best near the coast. Part shade in hot-summer climates. USDA: 8–10 Sunset: 6–9, 14–24 WUCOLS: L

^ *Arctostaphylos manzanita*

^ *Arctostaphylos* 'Monica'

A. 'Lester Rowntree', slow growing to 8–10 feet tall and 10–12 feet wide, with blue-gray leaves, new growth tipped with red-orange, and deep pink flowers. Natural hybrid involving *A. pajaroensis*, native along the coast and in the mountains of Monterey County, and possibly *A. obispoensis* from the Santa Lucia Mountains in San Luis Obispo County. USDA: 8–10 Sunset: N/A WUCOLS: L

A. manzanita, common manzanita, upright to 10–15 feet tall and 10–12 feet wide, with gray to bluish or dark green leaves and white or pink flowers. Native to rocky soils in open woodlands, grasslands, or chaparral in the inner North Coast Ranges and the northern Sierra Nevada foothills. Best away from the immediate coast. Not always easy in southern California, where *A. glauca* may be a better choice. USDA: 8–10 Sunset: 4–9, 14–24 WUCOLS: L/VL

A. 'Monica', 10–15 feet tall and wide, with glossy, dark green leaves and pink flowers. Natural hybrid from Sonoma County, California, believed to be between *A. densiflora* and *A. manzanita* and often offered as a selection of one or the other of these species. Adaptable. USDA: 8–10 Sunset: N/A WUCOLS: L

^ *Arctostaphylos* 'Pacific Mist'

A. 'Pacific Mist', sprawling, fast growing to 2–3 feet tall and 6–12 feet wide, with gray-green leaves, pink-tinged new growth, mahogany red stems upturned at the ends, and a few white flowers. Hybrid of garden origin, believed to involve *A. silvicola*, a gray-leaved plant from the mountains of coastal central California. Good for dryish shade. USDA: 7–10 Sunset: 7–9, 14–24 WUCOLS: L/M

A. uva-ursi, kinnikinnick or bearberry, less than 1 foot tall and 8–10 feet wide, with leathery, dark green leaves with bronzy or reddish tints in winter and pink or white flowers. Native to cool coastal or mountain habitats worldwide and in western North America from southeastern Alaska to coastal northern California. Many named cultivars. 'Point Reyes' is 12–18 inches tall and 6–8 feet wide. 'Radiant' and 'San Bruno Mountain' are prostrate mats 6–10 inches tall. Good choice for coastal gardens and the Pacific Northwest. USDA: 5–10 Sunset: 1–9, 14–24, A1–3 WUCOLS: L/M

^ *Arctostaphylos uva-ursi* 'Radiant'

Aristida purpurea

PURPLE THREE-AWN

Warm-season bunchgrass, upright to 1–2 feet tall and wide, with fine-textured, green to blue-green leaves and a haze of erect and arching, purple-tinted stems with reddish purple flowers in late summer or fall. Self-sows freely. Native to dry, rocky or sandy soils in the western two-thirds of the United States, southern Canada, and northern Mexico. Sun to part shade, good drainage, infrequent summer water. USDA: 6–10 Sunset: 3b–24 WUCOLS: L/VL

^ *Aristida purpurea*

Aristolochia californica

CALIFORNIA PIPEVINE

Deciduous vine, to 12–20 feet, with distinctive pipe-shaped, purple-striped, creamy white flowers on bare stems in late winter to early spring. Bright green, heart-shaped leaves are larval food for the pipevine swallowtail butterfly. Some old, brown leaves and seedpods hang on into the following growing season. For a more refined look, prune before new growth appears. Native to chaparral, woodlands, and forests below 1,500 feet, often along streambanks, in foothills and valleys of central and northern California. Part shade, or with roots in shade, most soils, occasional to infrequent summer water. USDA: 8–10 Sunset: 5–10, 14–24 WUCOLS: L/M

^ *Aristolochia californica*

Armeria maritima
SEA THRIFT, SEA PINK

Perennial, 4–6 inches tall and spreading slowly to 12–18 inches wide, with bright green, grasslike leaves and dense, rounded clusters of tiny bright pink or white flowers on leafless stems in spring or summer. Native to coastal dunes and bluffs in central and northern California, north to southeastern Alaska, and in coastal areas of Europe and the Mediterranean region. Cool sun, good drainage. Best along the immediate coast, where it needs infrequent to no summer water. Not low water in hot-summer climates. USDA: 4–8 Sunset: 1–9, 14–24, A2–3 WUCOLS: M

^ *Armeria maritima*

Artemisia
SAGEBRUSH

Evergreen shrubs and perennials with silvery gray, aromatic leaves and small, yellow or white flowers in summer. Grown for foliage. Sun to light shade, good drainage, infrequent to no summer water.

A. californica, California sagebrush, evergreen shrub, 2–4 feet tall and 4–6 feet wide, with narrow, feathery, gray-green leaves finely divided into threadlike lobes. Native to north- and west-facing slopes below 3,000 feet along the coast from central California to Baja California. 'Canyon Gray', from San Miguel Island, is less than 2 feet tall and 6–10 feet wide. Best near the coast but fine inland with good drainage and a little shade. USDA: 7–10 Sunset: 7–9, 14–24 WUCOLS: L/VL

^ *Artemisia californica* 'Canyon Gray'

A. 'Powis Castle', evergreen shrub, 2-3 feet tall and 4-6 feet wide, with flat sprays of finely divided leaves, green when new and maturing to silvery gray. Rarely flowers. Hybrid of garden origin, believed to be a cross between *A. arborescens*, native to the western Mediterranean region, and *A. absinthium*, more widely native to Europe, Asia, and northern Africa. USDA: 4-10 Sunset: 2-24 WUCOLS: L/M

A. pycnocephala, sandhill sage, evergreen shrub, 2-3 feet tall and slightly wider, with soft, pale green to gray or almost white, finely divided leaves. Native to rocky or sandy soils or beach sand along the immediate coast from Oregon to central California. 'David's Choice', from Point Reyes, California, is less than 1 foot tall. Best along the coast. USDA: 7-10 Sunset: 4-5, 7-9, 14-17, 19-24 WUCOLS: N/A

***A. schmidtiana* 'Nana'**, silver mound, perennial, 1 foot tall and 1-2 feet wide, with finely divided, silvery white leaves. Species is native to Japan. Dies to the ground in winter. May be offered as 'Silver Mound'. Best in cool-summer climates, where it needs occasional summer water. USDA: 3-8 Sunset: 1-10, 14-24, A1-3 WUCOLS: N/A

A. tridentata, big sagebrush, evergreen shrub, 3-6 feet tall and 4-8 feet wide, with strongly aromatic, blue-gray, wedge-shaped leaves covered with fine, silvery hairs. Flowers in long, loosely arranged clusters appear in late summer and fall. Native to a wide range of mountain habitats, from 1,500 to 10,000 feet, and from British Columbia south to Baja California and east to Texas. Infrequent to no summer water. Needs excellent drainage in high winter rainfall areas. May be offered as *Seriphidium*. USDA: 4-9 Sunset: 1-3, 6-11, 14-24 WUCOLS: VL

^ *Artemisia* 'Powis Castle'

^ *Artemisia tridentata*

Asclepias
MILKWEED

Perennials with narrow, green or gray-green leaves, loose clusters of star-shaped, summer flowers, and large, inflated pods containing seeds with silky hairs that aid in wind dispersal. Most die back to the ground in winter. Milkweeds spread, sometimes aggressively, by rhizomes and by seed. Leaves are larval food for butterflies. Sun to part shade, most well-drained soils, occasional to infrequent or no summer water.

A. californica, California milkweed, to 3 feet tall and wide, with broadly oval, woolly, gray-white leaves on woolly, white stems with pink to reddish purple flowers. Native to dry slopes and canyons in the Coast Ranges and Sierra Nevada foothills below 6,500 feet, most commonly in central and southern California and Baja California. USDA: 7–10 Sunset: N/A WUCOLS: L/VL

A. cordifolia, heartleaf milkweed, 1–3 feet tall and wide, with heart-shaped to broadly triangular, bluish green leaves and creamy white and deep purple flowers. Native to dry, rocky slopes and woodland openings from sea level to 6,500 feet in central and northern California, Oregon, and western Nevada. USDA: 7–10 Sunset: N/A WUCOLS: L/VL

A. fascicularis, narrowleaf milkweed, 2–3 feet tall and wide, with long, narrow, pointed, green leaves on upright stems topped by flat clusters of small, creamy white flowers tinted pink or lavender as they age. Can be weedy. Native to sunny, dry to moist habitats below 1,000 feet from southeastern Washington and Idaho to Oregon, California, and Baja California. USDA: 7–10 Sunset: N/A WUCOLS: L/VL

A. speciosa, showy milkweed, 3–4 feet tall and wide, with large, oval, softly hairy, gray-green leaves and pale pink or rose-purple flowers. Native to sunny, dry to moist habitats from sea level to 6,000 feet in much of western North America. USDA: 4–10 Sunset: 1–24 WUCOLS: N/A

^ *Asclepias speciosa*

^ *Asteriscus maritimus*

Asteriscus maritimus
GOLD COIN

Perennial, 1 foot tall and 3–4 feet wide, with small, spoon-shaped, green leaves and small, bright yellow daisy flowers in spring and summer. Native to the Mediterranean region along the coast from the Canary Islands to Greece. Full sun, good drainage, occasional to infrequent summer water. Best near the coast and in sandy soils. May be offered as *A.* 'Gold Coin' or as a *Pallenis* or *Odontospermum* species. USDA: 8b–11 Sunset: 9, 15–24 WUCOLS: L/M

Atriplex
SALTBUSH

Evergreen and deciduous shrubs with gray or silvery gray-green leaves and upright spikes of tiny summer flowers. Fruits sometimes have showy bracts. Native to desert, coastal, and riparian habitats nearly worldwide. Good habitat plants. May drop leaves in summer if grown dry. Sun, most soils, occasional to infrequent summer water. Cut back gently in late winter to renew. USDA: 6–10 WUCOLS: L/VL

A. canescens, fourwing saltbush, evergreen shrub, 2–6 feet tall and 4–8 feet wide, with narrow, linear, silvery gray-white, hairy leaves and yellow to red-brown flowers. Fruits on female plants are showy. Native to desert, intermountain, and Great Basin regions of the western United States. Sunset: 1–3, 7–24

A. lentiformis, big saltbush, evergreen shrub, 3–8 feet tall and 8–10 feet wide, with oval, almost triangular, silvery gray-green leaves and small yellowish summer flowers. Native to central California, east to southwestern Utah, and south into northern Mexico. The subspecies *breweri*, quail bush, has larger leaves and is native to coastal central and southern California and the Channel Islands. Sunset: 3, 7–14, 18–19

Baccharis
COYOTE BRUSH

Evergreen shrubs with small, resinous, glossy green leaves. Female plants have creamy white flowers with prominent, silky, white hairs that persist on the plant as seeds develop and aid in seed dispersal. Flowers on male plants are inconspicuous and do not produce seeds or the messy white floss associated with them. Sun, most soils, occasional to infrequent summer water. Trim annually or cut back hard every few years to renew.

< *Atriplex lentiformis* subsp. *breweri*

B. 'Centennial', 2–3 feet tall and 4–5 feet wide, is a female, seed-producing plant adapted to hot inland climates but thriving near the coast in full sun. *B.* THOMPSON ('Starn') is similar but does not produce seeds. Both are hybrids between *B. pilularis* and *B. sarothroides*, a species native to deserts of California and the Southwest. USDA: 8–10 Sunset: 7–24 WUCOLS: L/VL

B. magellanica, Christmas bush, a few inches tall and 4–6 feet wide, rooting as it spreads. Native to southern Argentina and southern Chile. Good seaside plant. USDA: 7–9 Sunset: N/A WUCOLS: N/A

B. pilularis, coyote brush, ground-hugging, mounding, or tall and upright, variable in the wild. Cultivated varieties are prostrate or low-mounding and widely spreading. Native to coastal dunes, dry slopes, and woodland openings in the Coast Ranges below 2,500 feet from northwestern Oregon to northwestern Baja California. 'Pigeon Point' and 'Twin Peaks 2', both 1–3 feet tall and 8 feet wide, are male plants that do not produce seed. Coastal plants, they need some shade inland. USDA: 7–10 Sunset: 5–11, 14–24 WUCOLS: L/M

^ *Baccharis pilularis*

^ *Baeckea virgata*

Baeckea
BAECKEA

Evergreen shrubs with fine-textured, needlelike, aromatic leaves and masses of small flowers in spring and summer. Cool sun, most well-drained soils, occasional to infrequent summer water. Sunset: N/A

B. gunniana, heath myrtle, to 2–3 feet tall and 2–4 feet wide, with midgreen leaves tinted coppery in winter and white flowers. Native to damp, rocky soils above 3,200 feet in the mountains of southeastern Australia from New South Wales to Tasmania. USDA: 7b–9 WUCOLS: N/A

B. virgata, tall baeckea, 8–10 feet tall and 6–8 feet wide, with midgreen leaves, new growth bronzy, and creamy white flowers. Native near the coast and in foothills of eastern Australia. Dwarf forms are sometimes available. Cut back to renew. May be offered as *Sannantha* or as *Babingtonia*. USDA: 8b–10 WUCOLS: L

Ballota pseudodictamnus
GRECIAN HOREHOUND

Perennial, mounding 1–2 feet tall and 3–4 feet wide, with aromatic, felted, gray-green leaves, gray-white beneath, and inconspicuous white flowers in summer. Native to rocky limestone soils in western Turkey and the Aegean Islands. 'All Hallows Green', long offered as a *Ballota* selection, is now considered a selection of *Marrubium bourgaei*. Sun, most well-drained soils, occasional to infrequent summer water. Accepts coastal winds and salt spray. USDA: 8–10 Sunset: N/A WUCOLS: VL

^ *Ballota pseudodictamnus* (foreground)

Banksia
BANKSIA

Evergreen shrubs and trees with narrow, often needlelike leaves and tiny flowers usually in large, cylindrical or conelike spikes. Sun to part shade, excellent drainage, good air circulation, occasional to infrequent summer water. USDA: 9–11 WUCOLS: L/M

B. ericifolia, heath banksia, 6–12 feet tall and wide, with needlelike, green leaves and red-orange flowers in fall and winter. Native to sandy soils in coastal southeastern Australia. Accepts sun or shade. Sunset: 15–24

B. marginata, silver banksia, variable, usually about 6 feet tall and wide, with dark green, needlelike leaves, silvery beneath, and yellow flowers in winter and spring. Native to a wide range of habitats in southeastern Australia. 'Mini Marginata' ('Mini-marg'), is 2 feet tall and frost hardy in mild-winter parts of the Pacific Northwest. Sunset: N/A

B. speciosa, showy banksia, 12–20 feet tall and 10–15 feet wide, with deeply serrated, olive-green leaves and creamy yellow flowers in summer and fall. Native to deep, sandy soils along the southern coast of Western Australia. Sunset: 15–24

^ *Banksia speciosa*

Berberis

BARBERRY, MAHONIA

Evergreen and deciduous shrubs with leaves divided into usually spiny-edged, glossy green to matte gray-green leaflets that may turn reddish or purple in winter. Bright yellow to yellow-orange flowers in spring are followed by red, blue, or purplish black berries. Cool sun to part shade, full shade in hot-summer climates, well-drained, humusy soils, moderate to occasional or infrequent summer water. May be offered as *Mahonia*. Some barberries, including *B. darwinii*, are considered invasive. All spread by seed, rhizomes, or rooting stems. Good for dryish shade.

B. aquifolium, Oregon grape, evergreen, upright to 4–6 feet tall and spreading, with glossy, dark green, spiny-edged leaflets, new growth coppery and winter accents bright red and purple. Native to foothills and mountains from British Columbia to northern and central California and at higher elevations in southern California. USDA: 5–10 Sunset: 2–12, 14–24 WUCOLS: M

B. aquifolium* var. *repens, creeping mahonia, evergreen, 1–3 feet tall and spreading, with dark blue-green leaves that turn reddish in winter. Native from British Columbia south to northern California and east to the Rocky Mountains. May be offered as *B. repens*. USDA: 4–10 Sunset: 2b–9, 14–24 WUCOLS: L/M

B. nevinii, Nevin's barberry, evergreen, 5–7 feet tall and spreading, with matte blue-green to dark gray-green leaves, new growth tinged pink. Native to sandy or gravelly soils in inland canyons and foothills of southern California. USDA: 5–10 Sunset: N/A WUCOLS: L/VL/M

^ *Berberis nevinii*

^ *Berberis thunbergii* 'Aurea'

B. thunbergii, Japanese barberry, deciduous, 4–6 feet tall and wide, with spiny branches and small green leaves that turn orange, yellow, and red in fall. Native to Japan and eastern Asia. Species is highly invasive in northeastern North America. 'Crimson Pygmy', with red leaves, and 'Aurea', with yellow leaves, produce little seed and are less likely to spread. USDA: 4–8 Sunset: 2b–24, A3 WUCOLS: L/M/H

Bergenia crassifolia

BERGENIA

Perennial, 1–2 feet tall and wide, spreading slowly by rhizomes, with large, leathery, dark green leaves and small, pink flowers on short stems in late winter or early spring. Native to rocky cliffs between 3,500 and 6,000 feet from Siberia south through northwestern China. Cool sun to shade, most soils, moderate to occasional or infrequent summer water. USDA: 4–8 Sunset: 1–9, 12–24, A1–3 WUCOLS: M/L

^ *Bergenia crassifolia*

Berlandiera lyrata

CHOCOLATE FLOWER

Perennial, 1–2 feet tall and wide, with gray-green leaves and chocolate-scented, night-blooming, yellow daisy flowers in spring and summer. Flowers are most fragrant in the early morning. May self-sow. Native to dry, gravelly or sandy soils from Colorado and Kansas south to Arizona, Texas, and northern Mexico. Full sun, most well-drained soils, occasional to infrequent summer water. Needs summer heat. USDA: 5–10 Sunset: 10–13, 18–23 WUCOLS: L/M

^ *Berlandiera lyrata*

Beschorneria yuccoides

AMOLE, MEXICAN LILY

Succulent rosette of gray-green, strap-shaped leaves, 3–4 feet tall and 4–5 feet wide, with pendant, tubular, bright green, late spring or summer flowers with prominent red bracts on bright red, 3- to 6-foot stalks. Clumps gradually expand by offsets at the base; mature rosette is persistent and repeat flowering. Native to rocky slopes in summer-rainfall regions of coastal southeastern Mexico. FLAMINGO GLOW ('Besys'), a selection from New Zealand, has variegated leaves; may be offered as 'Reality'. Full sun along the coast, part shade inland, most well-drained soils, occasional summer water. USDA: 8–11 Sunset: 13, 16–17, 19–24 WUCOLS: L/M

^ *Beschorneria yuccoides* in bloom

Bidens ferulifolia

BUR MARIGOLD

Perennial, 1 foot tall and 1–2 feet wide, with finely dissected, dark green leaves and masses of bright yellow daisy flowers in summer or almost year-round in mild-winter climates. May self-sow. Native to Arizona and New Mexico south through northern Mexico below 6,500 feet. Many cultivars. Full sun, well-drained soils, occasional summer water. USDA: 9–11 Sunset: 16–24 WUCOLS: L/M

^ *Bidens ferulifolia*

Bougainvillea

BOUGAINVILLEA

Evergreen or deciduous vines or vining shrubs, to 15–30 feet, with rounded to oval, matte green leaves on arching, spiny branches and inconspicuous, creamy white flowers surrounded by brightly colored, papery bracts in summer. Native to tropical and subtropical Central and South America. Many cultivars, the origins of which are often obscure. 'San Diego Red' is one of the more cold-hardy cultivars commonly available. Full sun, most well-drained soils, occasional to no summer water. USDA: 9b–11 Sunset: 12–17, 19–24 WUCOLS: L

^ *Bougainvillea* 'San Diego Red'

Bouteloua

GRAMA GRASS

Warm-season grasses, tufted, clump-forming, or slowly sod-forming with narrow, blue-green to grayish green leaves aging to tan, and distinctive flowerheads with flowers on one side of the stem in summer to fall. Sun to light shade, most well-drained soils, occasional to infrequent or no summer water. Good for erosion control. USDA: 4–9 Sunset: 1–3, 7–11, 14, 18–21

B. curtipendula, sideoats grama, 1–2 feet tall and wide, slowly sod-forming, with purplish flowers on 2- to 3-foot stems. Native to rocky, open slopes to 7,000 feet throughout much of North America. WUCOLS: L/VL

B. dactyloides, see Buchloe

B. gracilis, blue grama, 1 foot tall and 2 feet wide, clumping or sod-forming, with purple-tinged flowers on 2-foot stems. Native to the Great Plains region from Canada to the southwestern United States. 'Blonde Ambition' has chartreuse flowers on taller stems. Can be mowed as rough turf. May be offered as *Chondrosum*. WUCOLS: L

^ *Bouteloua gracilis* 'Blonde Ambition'

Brachyglottis greyi

DAISY BUSH

Evergreen shrub, slow growing to 3–5 feet tall and 4–6 feet wide, with felted, silvery gray-green leaves edged with white, silvery white beneath, and clusters of bright yellow daisy flowers in summer. Native to rocky, exposed bluffs on southern North Island, New Zealand, where rainfall is spread throughout the year. 'Silver Dormouse' is more compact with especially silvery leaves. Sun, good drainage, occasional summer water. Best along the coast. May be offered as *Senecio*. USDA: 8–9 Sunset: 5–9, 14–24 WUCOLS: L/M

^ *Brachyglottis greyi*

Brahea

HESPER PALM

Palms with stiff, flattened, fan-shaped leaves, stalks sometimes spiny, and long, pendant clusters of small white or yellowish flowers in late winter or early spring. Fruit drop can be messy. Sun, well-drained soils, infrequent to no summer water. WUCOLS: L

B. armata, blue hesper or Mexican blue palm, slow growing to 25–50 feet tall and 12–20 feet wide, with silvery gray-green leaves on spiny stalks, and long, pendant flower clusters that extend well below the leaves. Old leaves are persistent on the trunk. Native to rocky or sandy soils in desert canyons and washes in Baja California. Crown must be kept dry in summer. Needs summer heat. USDA: 8–11 Sunset: 12–17, 19–24

B. edulis, Guadalupe palm, slow growing to 20–30 feet tall and 12–15 feet wide, with medium green leaves. Drops old leaves. Native to dry, open woodlands on Guadalupe Island, off the coast of Baja California. Best near the coast. Does not do well in shade. USDA: 9–11 Sunset: 12–24

Brodiaea
BRODIAEA

Perennials from corms, with one to a few narrow, grasslike, basal leaves and a cluster of tubular to funnel-shaped flowers on a leafless stem in late spring or early summer. Deciduous and dormant after flowering. Spread slowly by offsets. Sun to part shade, well-drained clay soil, no summer water. Need moisture in spring to bloom well. USDA: 6–9 WUCOLS: L/VL

B. californica, California brodiaea, with lavender, white, or sometimes pink flowers. Native to valley grasslands and foothill woodlands below 5,000 feet in northern California and southwestern Oregon. Sunset: N/A

B. coronaria, early harvest brodiaea, with bluish or pinkish purple flowers. Native to seasonally moist places below 6,000 feet from southwestern British Columbia to central California, mostly west of the Cascades. The subspecies *rosea* has pink flowers. May be offered as *B. grandiflora*. Sunset: 4–9, 14–24

B. elegans, harvest brodiaea, with deep blue-purple flowers. Native to woodland openings and grassy meadows below 7,000 feet in mountains and foothills from southwestern Oregon to central California. Sunset: 2–9, 14–24

B. laxa, see *Triteleia*

B. pulchella, see *Dichelostemma*

< *Brahea edulis*

^ *Brodiaea elegans*

^ *Buchloe dactyloides*

Buchloe dactyloides
BUFFALO GRASS

Warm-season grass, 4–6 inches tall, with fine-textured, medium green to gray-green leaves that turn tawny brown in fall. Flowers in summer. Dormant in winter. Spreads by rhizomes to form a rough sod. Native to the shortgrass prairies of the Great Plains from central Canada to Texas. UC VERDE was developed for warmer climates of southern California. Sun, well-drained soils, occasional summer water. May be offered as *Bouteloua*. USDA: 4–9 Sunset: 1–3, 7–12, 15–16, 18–21 WUCOLS: N/A

^ *Bulbine frutescens* 'Hallmark'

Bulbine frutescens
STALKED BULBINE

Succulent rosette of narrow, grasslike leaves, 12–18 inches tall and wide, and bright yellow, star-shaped flowers on 2-foot stems in spring and summer. Spreads by rhizomes to form colonies 4–5 feet wide. Native to the southern coast of South Africa north to Mozambique. 'Hallmark' has orange flowers. Sun and infrequent to no summer water along the coast, afternoon shade and occasional water inland. Needs good drainage. May be offered as *B. caulescens*. USDA: 9–11 Sunset: 8–9, 12–24 WUCOLS: L

^ *Bulbinella nutans*

Bulbinella nutans
BULBINELLA

Perennial from rhizomes with a rosette of narrow, bright green, grasslike, somewhat fleshy leaves, 1 foot tall, slowly spreading to form colonies 4–5 feet wide. Conical clusters of small, bright yellow, star-shaped flowers on 3-foot stems in late winter or early spring. Native to southwestern Western Cape Province, South Africa. Deciduous and dormant in summer. Sun to light shade, well-drained soils, infrequent to no summer water. May be offered as *B. robusta*. USDA: 9–11 Sunset: N/A WUCOLS: L

^ *Bupleurum fruticosum*

Bupleurum fruticosum
SHRUBBY HARE'S EAR

Evergreen shrub, 4–5 feet tall and wide, with gray-white stems bearing narrowly oval, blue-green leaves, pale gray beneath, and loose, flat-topped to rounded clusters of chartreuse flowers in late spring and early summer. Native to rocky soils from Portugal and southern Spain east to Greece and south to coastal Morocco, Algeria, and Tunisia. Sun to light shade, good drainage, occasional to infrequent summer water. Good choice for exposed coastal sites. USDA: 7–10 Sunset: N/A WUCOLS: L

^ *Buxus sempervirens* 'Variegata'

Buxus sempervirens
BOXWOOD

Evergreen shrub, 5–15 feet tall and wide, with small, glossy, dark green leaves, paler green below, and inconspicuous flowers. Native to open woodlands and rocky hillsides, often near streams, from Portugal and Spain south to Morocco and east to Turkey. 'Graham Blandy', slow growing and narrowly upright to 5–10 feet tall and 1–2 feet wide, maintains its shape without pruning. 'Variegata', slow growing to about 5 feet tall, has green leaves with white margins that age to creamy yellow. Part shade or afternoon shade, good drainage, moderate to occasional summer water. Not low water in hot-summer climates. Does not do well in alkaline soils. USDA: 5–8 Sunset: 3b–6, 15–17 WUCOLS: M

Calamagrostis foliosa

MENDOCINO REEDGRASS

Cool-season bunchgrass, 1 foot tall and 2 feet wide, with narrow, gray-green leaves, purple tinged in winter, and silvery purple spring flowers on arching stems that age to pale gray-brown. Native to coastal bluffs and woodlands from sea level to 3,500 feet in northern California. Trim lightly or gently rake out old thatch; does not respond well to hard cutting back. Cool sun to part shade or afternoon shade, good drainage, occasional summer water. Best where both summers and winters are mild. Most other reedgrasses prefer regular water. USDA: 7–11 Sunset: N/A WUCOLS: M

^ *Calamagrostis foliosa*

Calandrinia spectabilis, see *Cistanthe grandiflora*

Calliandra

FAIRY DUSTER

Evergreen and semi-evergreen shrubs and perennials with small, finely divided, dark green or gray-green leaves and showy flowers with bristlelike stamens and no petals. Some species need regular summer water. Those listed here are content with occasional to no summer water in sun to part shade with good drainage. Drop leaves in summer if grown dry. USDA: 9–11 WUCOLS: L/VL

C. californica, Baja fairy duster, evergreen shrub, 4–6 feet tall and wide, with small, dark gray-green leaves and bright red flowers in spring and fall or almost year-round. Native to gravelly slopes and desert washes in Baja California. Infrequent to no summer water. Sunset: 10–24

C. eriophylla, pink fairy duster, semi-evergreen shrub, 2–3 feet tall and 3–4 feet wide, with gray-green leaves and pink flowers in late winter to spring. Native to sandy washes and dry, gravelly slopes from southeastern California east to western Texas and south to Baja California. Occasional summer water. Sunset: 11–24

^ *Calliandra californica*

Callistemon

BOTTLEBRUSH

Evergreen shrubs or small trees with narrow, lance-shaped, green to gray-green leaves and usually cylindrical flowerheads with bristly stamens in spring, summer, and sometimes fall. Native to summer-rainfall regions of eastern Australia, usually in moist habitats, but long grown in summer-dry climates with occasional or infrequent summer water. Sun to part shade, almost any soils. May be offered as *Melaleuca*.

C. 'Cane's Hybrid', 10–15 feet tall and wide, with grayish green leaves, new growth tinged pink, and pale pink and white flowers. Hybrid involving *C. pityoides or C. sieberi*. May be the same plant as 'Pink Stiletto' and 'Pink Ice'. USDA: 8b–11 Sunset: 8–9, 12–24 WUCOLS: L

C. *citrinus*, lemon bottlebrush, 8–20 feet tall and 6–15 feet wide, with green leaves, new growth coppery, and bright red flowers almost year-round. Native along the coast of southeastern Australia from Queensland to Victoria. 'Jeffers', 6 feet tall and 4 feet wide, has reddish purple flowers that age to lavender. May be offered as C. 'Jeffers' or as C. 'Violaceus'. USDA: 9–10 Sunset: 8–9, 12–24 WUCOLS: L

C. *pityoides*, alpine bottlebrush, variable, 3–8 feet tall and 4–6 feet wide, slowly forming dense thickets, with sharply pointed, green leaves and creamy yellow flowers. Native to the mountains of southeastern Australia. May be offered as a variety of C. *sieberi*. 'Corvallis', a selection from central western Oregon, may be especially hardy. Good choice for the Pacific Northwest. USDA: 7b–11 Sunset: 5–9, 14–24 WUCOLS: N/A

C. *viminalis*, weeping bottlebrush, fast growing to 15–20 feet tall and 15 feet wide, with light green leaves on pendulous branches and bright red flowers. Native along the eastern Australian coast and in foothills from northernmost Queensland to northeastern New South Wales. 'Little John' is 3 feet tall and wide. 'Captain Cook' is 6 feet tall and wide. SLIM ('CV01') is narrowly upright to 8–10 feet tall and 3–4 feet wide. USDA: 9–11 Sunset: 6–9, 12–24 WUCOLS: L/M

C. *viridiflorus*, green or mountain bottlebrush, upright to 5–7 feet tall and wide, with upward pointing, dark green leaves with maroon tints in winter and greenish yellow flowers. Native along streams in mountains of Tasmania. Needs some winter chill to flower well. USDA: 8–10 Sunset: 4–9, 14–24 WUCOLS: N/A

^ *Callistemon viminalis* 'Little John'

^ *Callistemon viminalis* SLIM

^ *Calluna vulgaris* 'Silver King'

Callitropsis, see *Hesperocyparis*

Calluna vulgaris

HEATHER

Evergreen shrub, variable, from low mats to upright shrubs, with tiny, scalelike, gray or green leaves that take on bronzy or purplish tones in winter and pink or pale purple, pendant, bell-shaped flowers in mid- to late summer. Spreads by seed and by rooting stems. Can be weedy. Native to open woodlands, scrublands, and grasslands from northern and western Europe to Siberia, Turkey, and Morocco. 'Firefly' has pinkish purple flowers and chartreuse leaves that age to orange and then to brick red. 'Silver King' has pinkish white flowers and pale blue-gray leaves. Cool sun to part shade, well-drained soils, moderate to occasional summer water. Not low water in hot-summer climates. USDA: 4–6 Sunset: 1a, 2–6, 15–17 WUCOLS: M

^ *Calluna vulgaris* 'Firefly'

Calocedrus decurrens

INCENSE CEDAR

Evergreen coniferous tree, slow growing to 60–80 feet tall and 15–20 feet wide, upright and pyramidal, with flat sprays of aromatic, dark green, scalelike leaves and reddish bark aging to gray. Native to shaded canyons and hot, dry mountain slopes, usually between 2,000 and 8,000 feet, from northern Oregon east to western Nevada and south through much of California to northern Baja California. Sun to light shade, good drainage, occasional to infrequent summer water. USDA: 5–8 Sunset: 2–12, 14–24 WUCOLS: M

^ *Calocedrus decurrens*

Calocephalus, see *Leucophyta*

Calochortus

MARIPOSA, GLOBE LILY, STAR TULIP

Perennials from bulbs that develop a few linear or lance-shaped leaves in winter, bloom in spring, and are dormant and deciduous in summer. Mariposas are tall plants with upward-facing, bowl-shaped flowers. Globe lilies, or fairy lanterns, have almost spherical, nodding flowers, usually several to a stem. Star tulips, or cat's ears, are small plants with upward-facing, bell-shaped flowers. Native to many summer-dry habitats in much of western North America. *C. superbus* is a usually white-flowered mariposa native to the North Coast Ranges and northern Sierra Nevada foothills. Sun to part shade, most well-drained soils. Must be kept dry in summer. USDA: N/A Sunset: N/A WUCOLS: L/VL

^ *Calochortus superbus*

Calothamnus quadrifidus

ONE-SIDED BOTTLEBRUSH

Evergreen shrub, 5–8 feet tall and wide, with needlelike, green to gray-green leaves and one-sided, brushlike spikes of red stamens in late winter and spring. Native to a wide range of habitats in southwestern Western Australia. Full sun, most well-drained soils, occasional to infrequent summer water. *C. villosus*, silky netbush, is similar, but not often available. USDA: 9b–11 Sunset: N/A WUCOLS: L

^ *Calothamnus quadrifidus*

Camassia

WILD HYACINTH, CAMAS

Perennials from bulbs, 1–2 feet tall and wide, with narrowly strap-shaped, green leaves and upright spikes of star-shaped flowers on leafless stems in spring. Summer dormant. Spread slowly by seed to form large colonies. Native to North America, especially the Pacific Northwest, in seasonally moist meadows that dry by late spring. Sun to part shade, well-drained humusy soils, no summer water. Must have ample water in spring to bloom well.

C. leichtlinii, large camas, with creamy white flowers on 3- to 4-foot stems. Subspecies *suksdorfii* has dark blue-violet or sometimes white flowers. Native west of the Cascades and on west-facing slopes of the Sierra Nevada below 8,000 feet, from southwestern British Columbia to northern California. USDA: 5–9 Sunset: 1–9, 14–17 WUCOLS: N/A

C. quamash, small camas, with bright blue to blue-violet flowers on 2- to 3-foot stems. Native from southwestern British Columbia south to California and east to western Montana, Wyoming, and Utah. USDA: 3–9 Sunset: 1–10, 14–17 WUCOLS: M

Camellia

CAMELLIA

Evergreen shrubs with leathery, glossy, dark green leaves, flowers in varied forms and colors, and growth habits ranging from sprawling to treelike. Native to summer-rainfall parts of China, Korea, and Japan. Numerous species and cultivars. *C. japonica* has unscented flowers in winter and early spring. Among the many cultivars of *C. japonica* are 'Alba Plena' with large, double, white flowers and 'Korean Fire' with single, almost funnel-shaped, deep red flowers. *C. sasanqua* has smaller leaves than *C. japonica* and smaller, sometimes fragrant flowers in fall and early winter. Among the many cultivars of *C. sasanqua* are 'Setsugekka' with semi-double, white flowers, 'Yuletide' with red flowers, and 'Kanjiro' with bright pink flowers. Cool sun to light shade, well-drained, moisture-retentive soils, moderate summer water. In southern California, plants may need regular summer water and shade. USDA: 7–10 Sunset: 4–9, 12, 14–24 WUCOLS: M

^ *Camassia quamash*

^ *Camellia sasanqua* 'Kanjiro'

^ Mixed *Carex* species

Carex
SEDGE

Perennials, clump-forming or spreading by rhizomes, with grasslike leaves and small, often inconspicuous flowers. Most are native to wetlands and other moist locations. Those listed here thrive with moderate to occasional summer water, especially near the coast. Most accept almost any soil in cool sun to light shade.

C. flacca, blue sedge, less than a foot tall and spreading vigorously by rhizomes to 2–3 feet wide, with blue-green leaves. Native to sandy soils in grasslands, marshes, and sand dunes of southern Europe and North Africa. 'Blue Zinger' has especially blue leaves. Occasional summer water. May be offered as *C. glauca*. USDA: 4–9 Sunset: 3–9, 14–24 WUCOLS: M/L

C. praegracilis, clustered field sedge, variable, less than a foot tall along the Pacific coast and up to 2 feet tall in inland mountain meadows, with narrow, glossy, green leaves. Native to much of western North America. Can be mowed as rough turf. Best with moderate water. Summer dormant if grown dry. *C. pansa*, dune sedge, is similar to coastal plants of *C. praegracilis*. USDA: 6–10 Sunset: 7–9, 11–24 WUCOLS: M

C. tumulicola, foothill sedge, to 1 foot tall and spreading slowly, with narrow, dark green leaves. Native to dry or moist meadows and open woodlands in the Coast Ranges from southern California to southwestern British Columbia and from sea level to 4,000 feet. Best right along the coast, where it is content in full sun or dryish part shade. USDA: 8–9 Sunset: N/A WUCOLS: L/M

Carpenteria californica

BUSH ANEMONE

Evergreen shrub, 6–8 feet tall and 3–5 feet wide, with glossy, dark green, lance-shaped leaves, gray-white and hairy beneath, and clusters of fragrant, white flowers in late spring and early summer. Native to dry woodlands, often along seasonal streams, in the Sierra Nevada foothills of Fresno County, California. 'Elizabeth' is more compact and has masses of smaller flowers. Full sun to part shade along the coast, shade or afternoon shade inland, most well-drained soils, occasional summer water. USDA: 8–10 Sunset: 5–9, 14–24 WUCOLS: M/L

^ *Carpenteria californica*

Ceanothus

WILD LILAC

Evergreen and deciduous shrubs, fast growing, mat-forming to medium-sized or tall, with glossy or matte, usually dark green, sometimes prickly leaves. Showy clusters of small, blue or white flowers, often from contrasting purplish to reddish pink buds, in winter to late spring. Sun, excellent drainage, infrequent to no summer water. Those listed here are evergreen.

C. arboreus, island or feltleaf ceanothus, 12–20 feet tall and 10–15 feet wide, with large, glossy, dark green leaves, felted and gray-white beneath, and pale blue flowers. Native to rocky slopes and exposed ridges on the Channel Islands in southern California. 'Powder Blue', from Santa Cruz Island, is 8–12 feet tall. Good small tree or tall, informal screen. USDA: 8b–10 Sunset: 5–9, 14–24 WUCOLS: L/M

C. 'Concha', 6–8 feet tall and 8–10 feet wide, with small, rough-textured, glossy, dark green leaves and dark blue flowers. Hybrid of garden origin believed to be between *C. impressus* and *C. papillosus* var. *roweanus*. Best near the coast but fine inland with some shade. Adaptable. USDA: 8–10 Sunset: 6–9, 14–24 WUCOLS: L

^ *Ceanothus* 'Concha'

C. *cuneatus*, buckbrush, variable, rounded or sprawling, 6–12 feet tall and wide, with small, leathery, wedge-shaped, dark green leaves and small clusters of fragrant, white to lavender or pale blue flowers. Native to coastal and interior mountains below 6,000 feet from north-central Oregon to northern Baja California. USDA: 7–9 Sunset: N/A WUCOLS: L/VL

C. 'Cynthia Postan', 6–8 feet tall and 6–10 feet wide, with small, glossy, dark green leaves and blue-violet flowers. Similar to C. 'Concha'. Hybrid of garden origin believed to be between *C. papillosus* var. *roweanus* and *C. thyrsiflorus* var. *griseus* from seeds collected in Berkeley, California, and grown in England. Performs well both in the Pacific Northwest and in southern California. USDA: 8–10 Sunset: 5–9, 14–24 WUCOLS: L

C. *gloriosus*, Point Reyes ceanothus, 1–5 feet tall and 6–10 feet wide, with glossy, dark green leaves with spiny margins and pale blue to dark blue-purple flowers. Native to seaside bluffs and dunes in coastal central California. Two low-growing selections of the variety *gloriosus* are 'Anchor Bay', 2–3 feet tall, and 'Heart's Desire', less than a foot tall, both 5–6 feet wide. *C. gloriosus* var. *porrectus*, Mt. Vision ceanothus, is about 1 foot tall and spreads more widely. USDA: 7b–10 Sunset: 4–6 WUCOLS: L/M

C. *griseus*, see C. *thyrsiflorus* var. *griseus*

C. 'Joyce Coulter', mounding to 2–4 feet tall and spreading 10–15 feet wide, with glossy, green leaves and medium blue flowers. Hybrid of garden origin believed to be between *C. thyrsiflorus* var. *griseus* and *C. papillosus*, a species from the Coast Ranges in central and southern California. Best in coastal locations but fine inland in part shade. Good bank cover. USDA: 8–10 Sunset: 5–9, 14–24 WUCOLS: L/M

C. 'Julia Phelps', 6–8 feet tall and 8–12 feet wide, with small, dark green, crinkled leaves and deep purple-blue flowers. Believed to be a hybrid of *C. impressus* and *C. papillosus* var. *roweanus*. Best in sun along the coast, where it needs no summer water. USDA: 8–10 Sunset: 5–9, 14–24 WUCOLS: L

^ *Ceanothus cuneatus*

^ *Ceanothus* 'Julia Phelps'

^ *Ceanothus maritimus* 'Valley Violet'

C. *maritimus*, maritime ceanothus, 1–3 feet tall and 4–8 feet wide, with small, leathery, dark green leaves and medium blue or blue and white flowers. Native to coastal bluffs in northern San Luis Obispo County, California. 'Valley Violet', with dark lavender-blue flowers, was selected for its good performance in the Sacramento Valley. USDA: 8–10 Sunset: 5–9, 14–24 WUCOLS: L

C. 'Midnight Magic', 3–4 feet tall and 5–6 feet wide, with glossy, dark green leaves and dark cobalt blue flowers. Hybrid between *C. papillosus* and *C. thyrsiflorus*. Good choice for small gardens. USDA: 8–10 Sunset: N/A WUCOLS: N/A

C. 'Ray Hartman', 10–20 feet tall and wide, with large, dark green leaves and medium blue flowers. Adaptable. Hybrid of *C. arboreus* and *C. thyrsiflorus* var. *griseus*. Best along the coast but fine inland with afternoon shade. Good small tree. USDA: 7b–10 Sunset: 5–9, 14–24 WUCOLS: L

C. *thyrsiflorus*, prostrate and widely spreading or upright and 15–20 feet tall, with glossy, dark green leaves, pale green beneath, and pale to dark blue or sometimes white flowers. Two varieties, with *C. thyrsiflorus* var. *thyrsiflorus* generally found at higher elevations than *C. thyrsiflorus* var. *griseus*. Prostrate and tall forms occur in both varieties, with differences between them mostly in the leaves.

< *Ceanothus* 'Ray Hartman'

C. *thyrsiflorus* var. *griseus*, Carmel ceanothus, variable, with blue to blue-purple flowers. Native to dunes, bluffs, and canyon slopes in coastal central and northern California. 'Kurt Zadnik', from the Sonoma County coast, with dark cobalt blue flowers, 'Yankee Point', from northern Monterey County, with bright blue flowers, and 'Hurricane Point', from the Big Sur coast, with pale blue flowers, are 2–3 feet tall and spread 10–15 feet wide. May be offered as 'Carmel Creeper' or as *C. griseus* var. *horizontalis*. Best along the coast. USDA: 8b–10 Sunset: 5–9, 14–17, 19–24 WUCOLS: L/M

C. *thyrsiflorus* var. *thyrsiflorus*, blueblossom, variable, with blue or sometimes white flowers. Native to coastal California from Santa Barbara north to the Oregon border. Some forms may be offered as the variety *repens*. 'Arroyo de la Cruz', from northern San Luis Obispo County, is 3–4 feet tall and 8–10 feet wide. Good bank cover. 'Snow Flurry', from the Big Sur coast in central California, is 10–15 feet tall and 12–20 feet wide, with brilliant white flowers. Part shade or afternoon shade inland. USDA: 8b–10 Sunset: 5–9, 14–24 WUCOLS: L

C. 'Victoria', 8–10 feet tall and wide, with dark green leaves and deep purplish blue flowers. Garden selection of uncertain parentage from Victoria, British Columbia, widely grown in the Pacific Northwest. Believed to be a hybrid between *C. thyrsiflorus* and *C. velutinus*, but often offered as a selection of *C. thyrsiflorus* or of *C. impressus*. May be the same plant as *C.* 'Skylark'. Adaptable. USDA: 7b–10 Sunset: 5–9, 14–24 WUCOLS: N/A

^ *Cercis canadensis* var. *texensis* 'Texas White'

^ *Cercis occidentalis*

Cercidium, see *Parkinsonia*

Cercis

REDBUD

Deciduous shrubs or small trees, usually multitrunk, with heart-shaped to rounded, green leaves, pink to magenta flowers in spring before leaves appear, and pendant seedpods that persist into winter. Sun to light shade, most well-drained soils, occasional to infrequent or no summer water. Need summer heat. Best in part shade.

C. canadensis* var. *texensis, Texas redbud, slow growing to 15–20 feet tall and 12–15 feet wide, with leathery, glossy, dark green, wavy-edged leaves, golden yellow or red in fall, pink flowers, and reddish purple seedpods. Native from southern Oklahoma through central Texas and into northeastern Mexico. 'Texas White' has white flowers and thicker leaves with slightly ruffled margins. May be offered as *C. reniformis*. USDA: 6–9 Sunset: 3–24 WUCOLS: M

C. occidentalis, western redbud, slow growing to 10–20 feet tall and 8–12 feet wide, with silvery gray bark, bright magenta flowers, purplish brown seedpods, and leaves that emerge lime green, mature to bluish green, and turn bright yellow in fall. Native to the Coast Ranges and Sierra Nevada foothills below 4,500 feet from northern to southern California and east to Arizona and Utah. Best in warm-summer climates, but flowers best with some winter chill. USDA: 7–9 Sunset: 2–24 WUCOLS: VL

Cercocarpus

MOUNTAIN MAHOGANY

Evergreen, deciduous, or semi-deciduous shrubs to small trees, narrowly upright and usually multitrunk, with leathery, dark green leaves and small clusters of inconspicuous, fragrant, creamy white to pale yellow flowers. Slow growing and long lived. Seeds have decorative, silvery white plumes that aid in dispersal by wind. Sun to light shade, most well-drained soils, occasional to infrequent or no summer water.

C. betuloides, mountain mahogany, evergreen, 10–15 feet tall and 6–10 feet wide, with small, deeply veined, oval to rounded leaves. Native to dry, rocky slopes below 6,000 feet in the mountains of southwestern Oregon to Baja California and east to central Arizona. May be offered as *C. montanus* var. *glaber*. USDA: 7–10 Sunset: 3, 5, 7–10, 13–24 WUCOLS: VL

C. ledifolius, curl-leaf or desert mountain mahogany, evergreen, slow growing to 6–25 feet tall and 3–10 feet wide, with lance-shaped leaves. Native to steep, rocky slopes between 1,200 and 3,000 feet from southeastern Washington to Baja California and from Montana to northwestern New Mexico. USDA: 5–10 Sunset: 1–3, 7–10, 14–21 WUCOLS: L/VL

^ *Cercocarpus betuloides*

^ *Cercocarpus ledifolius*

C. montanus, alderleaf mountain mahogany, deciduous or semi-deciduous, 4–12 feet tall and 6–12 feet wide, with oval to oblong leaves that turn deep yellow in fall. Native to dry, rocky slopes and mesas, usually between 3,000 and 8,000 feet, from southwestern Oregon to Baja California and from southern Montana to Texas and northern Mexico. USDA: 4–10 Sunset: 1–3, 7–10 WUCOLS: VL

Chaenomeles

FLOWERING QUINCE

Deciduous shrubs, 3–10 feet tall and slowly thicket forming by suckering roots, with dark green leaves on usually spiny branches and showy flowers in late winter to early spring. Native to China, Japan, Korea, and Myanmar. Several species and hundreds of cultivars. *C.* ×*superba* 'Pink Lady' is a thornless cultivar, 5 feet tall and wide, with deep rose pink flowers. Sun to part shade, most soils, occasional summer water. USDA: 4–8 Sunset: 2–23 WUCOLS: L/M

^ *Chaenomeles* ×*superba* 'Pink Lady'

^ *Chamaerops humilis*

^ *Chamelaucium ciliatum* 'Scaddan'

Chamaerops humilis

MEDITERRANEAN FAN PALM

Palm, slow growing and shrubby to 8–10 feet tall and 10–15 feet wide, often forming multiple trunks with age. Leaves consist of 10–20 radiating, green to bluish green leaflets on 3- to 4-foot stalks covered in sharp, needlelike spines. Bright yellow flowers and orange fruits are mostly hidden among the leaves. Native to Mediterranean coasts from Spain to Italy and from Morocco to Tunisia. The variety *argentea*, from the Atlas Mountains, has silvery blue leaves; may be offered as variety *cerifera*. 'Vulcano' has no spines. Full sun, most well-drained soils, occasional summer water. USDA: 8–11 Sunset: 4–24 WUCOLS: L

Chamelaucium

WAXFLOWER

Evergreen shrubs with aromatic, needlelike leaves and small flowers in late winter and spring or for much of the year. Native to sandy soils in southwestern Western Australia. Sun to part shade, fast-draining soils, infrequent to no summer water. USDA: 9–11 Sunset: 8–9, 12–24 WUCOLS: L

C. ciliatum, Stirling waxflower, variable, 3–4 feet tall and wide, with gray-green leaves and white flowers aging to pink. 'Scaddan' is 3 feet tall and 3–4 feet wide, with pinkish white flowers that age to deep rose pink.

C. uncinatum, Geraldton waxflower, 6–10 feet tall and wide, with green leaves and small pink, purplish, or white flowers aging to pink. 'Purple Pride', 4–6 feet tall and wide, has dark pinkish purple flowers. 'Dancing Queen' has pale lilac double flowers.

^ *Chilopsis linearis*

Chilopsis linearis
DESERT WILLOW

Deciduous shrub to small tree, often multitrunk, fast-growing to 15–30 feet tall and 10–20 feet wide, with long, narrow, green to gray-green leaves and clusters of fragrant, trumpet-shaped, pink to white or purplish flowers from spring to fall. Long, narrow seedpods hang from the tree into winter. Native to sandy or rocky soils in desert washes below 5,000 feet from southern California to northern Mexico and east to western Texas. ART'S SEEDLESS, 'Lois Adams', and TIMELESS BEAUTY ('Monhews') do not produce seedpods. 'Burgundy' has gray-green leaves and deep purple flowers. Best in hot-summer, mild-winter climates. Full sun, good drainage, occasional to infrequent summer water. USDA: 7b–11 Sunset: 3b, 7–14, 18–23 WUCOLS: L/VL

^ ×*Chitalpa tashkentensis* 'Morning Cloud'

×*Chitalpa tashkentensis*
CHITALPA

Deciduous tree, fast growing to 20–30 feet tall and wide, usually multitrunk, with dark green, lance-shaped leaves and upright clusters of funnel-shaped, wavy-edged, white, pink, or lavender flowers from late spring into fall. Hybrid between *Chilopsis linearis*, native to the deserts of California, Texas, and northern Mexico, and *Catalpa bignonioides*, native from Mississippi to Georgia and northwestern Florida. 'Pink Dawn' and SUMMER BELLS ('Minsum') have pale pink flowers. 'Morning Cloud' has white flowers. Full sun, most well-drained soils, occasional to infrequent or no summer water. USDA: 6–9 Sunset: 3–24 WUCOLS: N/A

Choisya ternata
MEXICAN ORANGE

Evergreen shrub, 5–8 feet tall and wide, with glossy, bright green leaves and clusters of fragrant, pure white flowers in early spring to summer. May spread by underground runners to form a thicket. Native to the mountains of central Mexico. SUNDANCE ('Lich'), with bright yellow new leaves, is best with afternoon shade. *C.* ×*dewitteana* 'Aztec Pearl', a hybrid between *C. ternata* and *C. dumosa* var. *arizonica*, from southeastern Arizona and northwestern Mexico, has narrower leaflets, larger flowers, and a more upright habit. Cool sun to light shade, most well-drained soils, occasional to infrequent summer water. USDA: 8–10 Sunset: 6–9, 14–24 WUCOLS: M

^ *Choisya ternata* SUNDANCE

Chondropetalum, see *Elegia*

Chondrosum, see *Bouteloua*

Chrysanthemum, see *Rhodanthemum*

Cistanthe grandiflora
ROCK PURSLANE

Perennial, 8–10 inches tall and spreading 3 feet wide, rosette of succulent, bluish gray-green leaves and magenta flowers on 2- to 3-foot stems from spring to fall. Native to the mountains of central Chile. Sun to part shade, well-drained soil, occasional to infrequent summer water. May be offered as *Calandrinia*. USDA: 8–11 Sunset: 15–17, 20–24 WUCOLS: L

^ *Cistanthe grandiflora*

^ *Cistus ×purpureus*

Cistus
ROCKROSE

Evergreen shrubs with rough-surfaced, dark green to gray-green, often aromatic leaves and showy, white or pink flowers with bright yellow stamens in spring to midsummer. Flowers drop petals in late afternoon and new buds open by morning. Native to rocky soils throughout the Mediterranean region from the Canary Islands to the Caucasus and from Morocco to Jordan and Israel. Dozens of cultivars. *C. creticus*, *C. ladanifer*, and *C. monspeliensis* are considered invasive or potentially so, and all should be watched for any tendency to spread. Sun to part shade or afternoon shade, fast drainage, infrequent to no summer water. Tip prune only as needed; rockroses do not respond well to hard cutting back. USDA: 8–10 Sunset: 4–9, 14–24 WUCOLS: L/M

C. ×*aguilarii*, upright to 5–7 feet tall and 4–6 feet wide, with dark green leaves and white flowers. Natural hybrid involving *C. ladanifer* and *C. populifolius*, a species with bright green, wavy-edged leaves and white flowers. Often offered as *C.* 'Blanche', a cultivar involving *C. ladanifer* and possibly a cross with *C. palinhae*. *C. ×aguilarii* 'Maculatus' has purplish blotches at the base of each petal.

C. ×*bornetianus* 'Jester', rounded and upright to 2–3 feet tall and 4 feet wide, with silvery gray-green leaves and soft pink flowers. Hybrid between *C. albidus* and *C. laurifolius*.

C. ×*hybridus*, white rockrose, 2–4 feet tall and 6–10 feet wide, with dark gray-green leaves and pure white flowers with yellow spots at the petal bases. Natural hybrid of *C. salviifolius* and *C. populifolius* discovered in coastal southern Europe. May be offered as *C. ×corbariensis*.

C. ×*purpureus*, orchid rockrose, 4–6 feet tall and 6–8 feet wide, with dark green leaves and magenta flowers with a reddish brown blotch at the base of each petal. Hybrid between *C. ladanifer* and *C. creticus*.

C. salviifolius, sageleaf rockrose, 2–3 feet tall and 6–8 feet wide, with gray-green leaves and white flowers with a yellow spot at the base of each petal. Usually available as 'Prostratus'.

^ *Clarkia concinna*

^ *Clarkia unguiculata*

^ *Clematis lasiantha*

Clarkia

CLARKIA

Annuals with narrow, linear to lance-shaped leaves and cup-shaped or broadly open and lobed flowers in late spring and early summer. Self-sow. Sun to light shade or afternoon shade, most well-drained soils, occasional summer water. Best from seed sown in open ground in fall. Need moisture in spring to bloom well. USDA: N/A Sunset: 1–24, A2–3 WUCOLS: N/A

C. amoena, farewell-to-spring, 1–3 feet tall and 2 feet wide, with cup-shaped, pink to lavender or pale purple flowers, sometimes with darker blotches at the petal bases. Native to coastal scrub, grassland, and woodland or forest openings below 3,000 feet from British Columbia south to the San Francisco Bay Area. May be offered as *Godetia*, the name commonly used in the cut flower trade.

C. concinna, red ribbons, 1 foot tall and 2 feet wide, with deep pink to red flowers with trilobed, fan-shaped petals. Native to foothills and low mountains in central and northern California from sea level to 5,000 feet. Prefers part shade or afternoon shade.

C. unguiculata, elegant clarkia, 1–3 feet tall and less than 1 foot wide, with pink to reddish purple flowers with paddle-shaped petals on tall stems. Native to the central California coast and northern Sierra Nevada foothills.

Clematis lasiantha

PIPESTEM OR CHAPARRAL CLEMATIS

Deciduous vine, to 15 feet, with medium green leaves divided into several lobed leaflets and fragrant, white flowers with prominent yellow stamens in winter and spring. Showy seedheads with long, feathery tails. Native to partly shaded, rocky slopes below 6,000 feet in the Coast Ranges and western Sierra Nevada foothills from the San Francisco Bay Area to Baja California. Cool sun to part shade, most well-drained soils, infrequent to no summer water. USDA: 5–10 Sunset: N/A WUCOLS: L/VL

^ *Comarostaphylis diversifolia*

^ *Condea emoryi*

Comarostaphylis diversifolia

SUMMER HOLLY

Evergreen shrub, upright and slow growing to 8–20 feet tall and 5–8 feet wide, with shredding gray bark exposing reddish brown underbark and glossy, leathery, dark green leaves with finely serrated margins. Pendant clusters of small, urn-shaped, white flowers in spring followed by bright red berries. Native to shaded, dry slopes, often near seasonal streams, in coastal southern California, including the Channel Islands, and northern Baja California. Part shade or afternoon shade, most well-drained soils, no summer water. USDA: 8–10 Sunset: 7–9, 14–24 WUCOLS: L/VL

Condea emoryi

DESERT LAVENDER

Evergreen shrub, 4–8 feet tall and wide, with silvery gray, woolly leaves and lavender flowers in early spring. Native to sandy soils on hot, dry slopes from southeastern California to Arizona and northwestern Mexico. Sun to part shade, excellent drainage, infrequent summer water. Needs summer heat. May be offered as *Hyptis*. USDA: 9–11 Sunset: 8–14, 18–24 WUCOLS: L/VL

^ *Constancea nevinii*

Constancea nevinii

NEVIN'S WOOLLY SUNFLOWER

Evergreen shrub, 3–6 feet tall and wide, with finely divided, silvery gray, almost white leaves and tight clusters of tiny, golden yellow, early summer flowers that age to a rich, dark brown. Native to rocky coastal bluffs on the southern Channel Islands. Full sun to part shade, most well-drained soils, infrequent to no summer water. Best near the coast. Cut back in late fall or winter to maintain dense form. May be offered as *Eriophyllum*. USDA: 9b–11 Sunset: 15–17, 19–24 WUCOLS: L/VL

Convolvulus cneorum

BUSH MORNING GLORY

Evergreen shrub, fast growing to 2–3 feet tall and 3–4 feet wide, with silvery green, lance-shaped leaves and funnel-shaped, white flowers from pink buds in late spring to fall. Native to rocky soils along the coast from Spain to Albania. Sun to light shade, good drainage, occasional summer water. Afternoon shade in hot-summer climates. Cut back in late winter to renew. May be short lived. USDA: 8–10 Sunset: 7–9, 12–24 WUCOLS: L

Convolvulus cneorum >

Coreopsis gigantea, see *Leptosyne*

Corethrogyne, see *Lessingia*

Correa

AUSTRALIAN FUCHSIA

Evergreen shrubs with small, oval to rounded, dark green to gray-green leaves, paler beneath, and tubular flowers with flared tips, usually in fall or winter to spring. Native to southeastern Australia. Most plants offered are cultivars. Cool sun to part shade, fast drainage, occasional to infrequent summer water. Best near the coast. USDA: 9–10 Sunset: 14–24 WUCOLS: L

C. *alba*, white correa, 6–8 feet tall and wide, with dark green leaves and white flowers.

C. *backhouseana*, 4–5 feet tall and wide, with gray-green leaves and pink-tinged, creamy white or greenish white flowers.

C. 'Dawn in Santa Cruz', 3–5 feet tall and 5–8 feet wide, with gray-green leaves and pink flowers with pale yellow tips.

C. 'Dusky Bells', red Australian fuchsia, 2–3 feet tall and 4–6 feet wide, with waxy, dark green leaves and red to reddish pink flowers. Believed to be a hybrid between *C. reflexa* and *C. pulchella*. Best in part shade. May be offered as 'Carmine Bells'.

C. 'Ivory Bells', 4–5 feet tall and wide, with dark gray-green leaves on coppery stems and creamy white flowers. Hybrid between *C. alba* and *C. backhouseana*.

C. *pulchella*, pink Australian fuchsia, 2–3 feet tall and 5–6 feet wide, with smooth, shiny, dark green leaves and light pink flowers. 'Pink Flamingo', a selection from coastal central California, has deep salmon-pink flowers.

^ *Correa* 'Dawn in Santa Cruz'

C. *reflexa*, variable, from open and upright to low and spreading, with leaves smooth or rough and hairy and flowers greenish yellow to deep red. 'Cape Carpet' is less than 1 foot tall and 8–10 feet wide with small, bright red flowers with chartreuse petal tips. 'Kangaroo Island', 3–4 feet tall and wide, has red flowers.

^ *Corymbia ficifolia*

Corymbia
GUM TREE

Evergreen trees with thick, leathery, aromatic leaves and sometimes showy flowers. Native to Australia. Sun to part shade, well-drained soils, occasional to infrequent summer water. Good coastal trees. May be offered as *Eucalyptus*. Sunset: 5–6, 8–24 WUCOLS: L/M

C. citriodora, lemon-scented gum, 60–90 feet tall and 25–40 feet wide, with smooth, white, peeling bark, yellow-green leaves, and small clusters of white flowers in winter. Native to dryish woodlands in the mountains and along the coast from northern Queensland to New South Wales. USDA: 9–11

C. ficifolia, red-flowering gum, 25–40 feet tall and wide, with rough, fissured, black-brown bark, dark green leaves on somewhat pendulous branches and showy clusters of red or sometimes pink flowers in late summer. Native to open forest on the southwestern coast of Western Australia. USDA: 10–11

Cotinus
SMOKE TREE

Deciduous shrubs to small trees, upright and multistem, with large, broadly oval, green to bluish green or sometimes purple leaves that turn yellow,

^ *Cotinus* 'Grace'

orange, or red in fall. Clusters of tiny, creamy white flowers in spring, followed by a billowy haze of showy, pale pink to deep purple silky hairs as the flowers age. Full sun, most soils, moderate to occasional summer water. USDA: 5–8 Sunset: 2–24 WUCOLS: L

C. *coggygria*, smoke tree, 12–15 feet tall and wide, with bluish green leaves. Native from southern Europe across central Asia and the Himalayas to northern China. 'Royal Purple' and 'Velvet Cloak' have deep purple leaves.

C. 'Grace', 12–20 feet tall and wide, with red new leaves that age to dark red or purple. Hybrid of *C. coggygria* 'Velvet Cloak' and *C. obovatus*.

C. *obovatus*, American smoke tree, 15–30 feet tall and 12–20 feet wide, with bluish to dark green leaves. Native to the southeastern United States from Oklahoma to Tennessee and south to Alabama.

Cotyledon orbiculata
PIG'S EAR

Succulent, 2–3 feet tall and wide, with rounded, gray-white leaves lightly edged in red and tall stems topped with clusters of bell-shaped, pale orange, summer flowers. Native to dry, rocky cliffs and ridges in Africa, especially South Africa. The variety *oblonga* has fingerlike, gray-green leaves. 'White Platter' has especially large, nearly white leaves. Cool sun to light shade, good drainage, occasional to infrequent or no summer water. Not hardy inland in northern California. USDA: 10–11 Sunset: 12–13, 16–17, 21–24 WUCOLS: L

Crassula
CRASSULA

Succulents with thick, fleshy leaves and clusters of small, star-shaped flowers. Most of the hundreds of plants offered are cultivars of species native to South Africa. Cool sun to light shade, good drainage, occasional to infrequent or no summer water. Not hardy inland in northern California. USDA: 10–11 Sunset: 8–9, 12–24 WUCOLS: L

C. multicava, fairy crassula, to 1 foot tall and wide, with glossy, dark green, rounded leaves and showy sprays of white flowers in winter. Native to forest edges and along the southeast coast of South Africa from KwaZulu-Natal to the Western Cape. Good for dry shade.

C. ovata, jade plant, 4–6 feet tall and wide, forms a thick trunk and multiple branches with oval, green leaves lightly edged in red and clusters of white or pinkish white flowers in winter. Native to rocky outcrops in Eastern Cape Province. 'Gollum', slow growing to 1–2 feet tall, has tubular leaves. May be offered as *C. argentea* or as *Cotyledon*.

C. perfoliata, 1–2 feet tall and wide, with long, gray-green, sickle-shaped leaves and showy, rounded to flat-topped clusters of red or pink flowers in midsummer. The variety *falcata*, airplane plant, with bright red flowers, may be offered as variety *minor* or as *C. falcata*. Native along the coast and inland in southwestern Eastern Cape Province.

^ *Cotyledon orbiculata*

^ *Crassula multicava*

^ *Cyclamen hederifolium*

Cupresses arizonica, C. forbesii, C. sargentii, see *Hesperocyparis*

Cupressus sempervirens

ITALIAN OR MEDITERRANEAN CYPRESS

Evergreen coniferous tree, variable, densely upright or more open and spreading, 40–60 feet tall and 8–12 feet wide, with dark green, aromatic, scalelike leaves and small, round cones on upright branches. Native to mountains of the eastern Mediterranean region from Greece to Turkey and from Libya to Lebanon. 'Glauca' has blue-green leaves. 'Totem Pole' and TINY TOWER ('Monshel') are dwarf selections. Full sun, most well-drained soils, infrequent to no summer water. USDA: 7–10 Sunset: 4–24 WUCOLS: L/M

< *Cupressus sempervirens* 'Glauca'

Cyclamen

CYCLAMEN

Perennials from tubers, with rounded to heart-shaped or lobed leaves and nodding flowers on leafless stems. Summer dormant. Self-sow. Part shade, humusy, fast-draining soils, infrequent to no summer water. Sunset: 2–9, 14–24

C. coum, hardy cyclamen, 2–3 inches tall and 6 inches wide, with dark green leaves that appear in fall, sometimes patterned with silver or gray, and white or pink to magenta flowers in winter or early spring. Native to open woodlands from sea level to 7,000 feet, from Turkey to Israel and around the Black Sea. Prefers some winter chill. USDA: 6–9 WUCOLS: L

C. hederifolium, ivy-leaved cyclamen, 4–6 inches tall and 6 inches wide, with dark green leaves, maroon beneath, patterned with pale green, silver, or gray and small pink or sometimes white flowers. Flowers appear with fall rains, followed by leaves that persist through winter and die back in spring. Native to the Mediterranean region from sea level to 4,000 feet and from southern France to Turkey. USDA: 5–9 WUCOLS: L/VL

Dalea
DALEA

Evergreen shrubs and perennials with fine-textured leaves divided into small leaflets and spikes of tiny purple flowers. Full sun, well-drained soils, occasional to infrequent or no summer water. Need summer heat. WUCOLS: L

D. frutescens, black dalea, evergreen shrub, fast growing to 3–4 feet tall and wide, with silvery gray-green leaves and flowers in winter and spring. Native to dry, rocky slopes in deserts of Texas, New Mexico, and northern Mexico. Drops leaves in summer if grown dry. USDA: 8–11 Sunset: 10–13

D. greggii, trailing indigo bush, evergreen shrub, fast growing to 1 foot tall and 3–6 feet wide, dense and mounding, with silvery blue-green leaves and flowers in spring and summer. Cascades over walls. Native to New Mexico, Texas, and northern Mexico from 2,000 to 4,500 feet. Good groundcover for southern California. USDA: 8–11 Sunset: 10–13

D. purpurea, purple prairie clover, perennial, 1–3 feet tall and 1–2 feet wide, with dark green leaves on upright stems and flowers in summer. Self sows. Native to rocky soils in open woodlands and grasslands in the Great Plains region from Canada to New Mexico. May be offered as *Petalostemon*. USDA: 4–8 Sunset: 1–3, 10–12

^ *Dalea purpurea*

^ *Dasylirion wheeleri*

Dasylirion
SOTOL

Succulent rosettes of stiff, narrow, grasslike leaves, some with sharp marginal teeth and some slowly forming a short or tall stem. All eventually develop flowering stalks, 10–15 feet tall, and flower repeatedly but not every year. Sun to light shade, well-drained soils, occasional to infrequent or no summer water. USDA: 8–11 Sunset: 10–24 WUCOLS: L/VL

D. longissimum, Mexican grass tree, 4–6 feet tall and 6–10 feet wide, with long, bluish gray-green leaves without teeth and white flowers from reddish buds. Slowly forms a 6- to 15-foot stem. Native to deserts of northeastern Mexico. May be offered as *D. quadrangulatum*.

D. texanum, green desert spoon, 3–5 feet tall and wide, with light green leaves with sharp marginal teeth and greenish yellow flowers. Slowly develops a short woody stem, sometimes partly underground. Native to western Texas and northeastern Mexico.

D. wheeleri, desert spoon, 3–5 feet tall and 3–6 feet wide, with blue-gray to gray-green leaves with sharp marginal teeth and greenish white to pale pink flowers. Slowly develops a 4- to 6-foot stem. Native to rocky slopes and grasslands from 3,000 to 6,000 feet in southeastern Arizona, southwestern New Mexico, and western Texas to northeastern Mexico.

Delosperma

ICE PLANT

Succulents with tiny, cylindrical, green to bluish green leaves and bright pink to purple daisy flowers. Full sun, excellent drainage, occasional summer water. Good alternative to invasive *Carpobrotus* species. Sunset: 2–24 WUCOLS: L

D. cooperi, purple ice plant or pink carpet, 3–4 inches tall and 1–2 feet wide, with green leaves and purple flowers. Native to summer-rainfall regions of South Africa. FIRE SPINNER ('P001S') has tricolor flowers of pink, red, and orange. May be offered as *Mesembryanthemum*. USDA: 7–10

D. floribundum, hardy ice plant, 4–6 inches tall and 10–18 inches wide, with bluish green leaves and purple flowers with white centers. Native to grassy ridges and plains of Free State, South Africa, where summers are hot and rainy and winters are cold and dry. Often available as STARBURST. USDA: 6–9

^ *Delosperma floribundum*

Dendromecon

TREE POPPY

Evergreen shrubs, fast growing, with thick, oval to lance-shaped, bluish gray-green leaves and bright yellow flowers in spring and early summer, followed by long seedpods. Best along the coast. Not always easy inland. Sun to part shade, excellent drainage, infrequent to no summer water.

D. harfordii, island tree poppy, 6–12 feet tall and wide. Native to dry slopes in chaparral below 2,000 feet on the Channel Islands in southern California. May be offered as a subspecies of *D. rigida*. USDA: 7–10 Sunset: 7–9, 14–24 WUCOLS: L/VL

D. rigida, bush poppy, 4–8 feet tall and 5–6 feet wide, with narrower leaves and a more open habit than *D. harfordii*. Native to dry slopes below 6,000 feet in the Coast Ranges and Sierra Nevada foothills from central California to northern Baja California. USDA: 6–10 Sunset: 4–12, 14–24 WUCOLS: VL

^ *Dendromecon harfordii*

Dichelostemma congestum

Dicliptera sericea

Dichelostemma

WILD HYACINTH

Perennials from corms, with several grasslike leaves in late winter or early spring, followed by dense, rounded clusters of tubular to bell-shaped flowers on tall stems. Deciduous and dormant in summer. Spread slowly by offsets. Sun to part shade, most well-drained soils, infrequent to no summer water. USDA: 6–9 Sunset: N/A WUCOLS: VL

D. capitatum, blue dicks, with blue-violet flowers, sometimes purplish pink or white. Native to grasslands and open woodlands from sea level to 7,000 feet in western Oregon, much of California, and east to Utah, New Mexico, and northwestern Mexico. May be offered as a variety of *D. pulchellum* or as *Brodiaea*.

D. congestum, ookow, with purplish blue flowers. Native to meadows and woodland openings below 6,000 feet from southwestern British Columbia to central California. May be offered as *D. pulchellum* or as *Brodiaea pulchella*.

D. ida-maia, firecracker flower, with pendant, purplish red and creamy white flowers. Native to meadows, woodlands, and forest openings near the coast below 6,000 feet in northwestern California and southwestern Oregon. May be offered as *Brodiaea*.

Dicliptera sericea

FIRECRACKER PLANT

Perennial, 1–2 feet tall and wide, with velvety, bluish gray stems and leaves and clusters of upright, bright red-orange, tubular flowers in summer and fall. Dies back to the ground in winter. Native to Uruguay, Paraguay, and parts of Brazil with year-round distribution of rainfall. Full sun, most soils, moderate to occasional summer water. May be offered as *D. suberecta*. USDA: 8–10 Sunset: 12–13, 15–17, 19, 21–24 WUCOLS: L

Diplacus aurantiacus
STICKY MONKEYFLOWER

Perennial or subshrub, 1–3 feet tall and wide, with glossy, dark green, resinous leaves and orange or yellow, sometimes red or creamy white flowers. Native to rocky or sandy soils in many habitats from southwestern Oregon to Baja California. Dormant and semi-deciduous in midsummer where summers are hot, perking up again with fall rains. Most plants offered are selections or hybrids of six naturally occurring varieties that differ in several flower and leaf characteristics. Sun or part shade, most well-drained soils, infrequent summer water. Hybrids may need more water. May be short lived in heavy soils. May be offered as *Mimulus*. USDA: 7–10 Sunset: 7–9, 14–24 WUCOLS: L/VL

^ *Diplacus aurantiacus*

Dodonaea viscosa
HOPBUSH

Evergreen shrub, fast growing to 10–15 feet tall and 8–12 feet wide, with long, narrow, green to purplish green leaves, inconspicuous flowers, and showy, long-lasting, winged seedpods. May spread by seed. Native to a wide range of climates and soils in much of Australia, New Zealand, and parts of the southwestern United States and northeastern Mexico. 'Purpurea' has bronzy green leaves that turn deep purple in cooler weather. Sun to light shade, most well-drained soils, infrequent to no summer water. USDA: 9–11 Sunset: 7–24 WUCOLS: L/M

^ *Dodonaea viscosa* 'Purpurea'

Dorycnium hirsutum
CANARY CLOVER

Evergreen shrub, 1–3 feet tall and wide, with softly hairy, gray-green leaves and small white flowers from pink buds in summer to fall. Native to grasslands of the Mediterranean region from Portugal to Turkey. Full sun, good drainage, occasional to infrequent or no summer water. Good seaside plant. Fairly short lived. USDA: 8–10 Sunset: N/A WUCOLS: L/M

^ *Dorycnium hirsutum*

^ *Dryopteris arguta*

Dryopteris arguta

COASTAL WOOD FERN

Fern, 1–3 feet tall and wide, with erect, dark green fronds. Native to north-facing slopes and shady woodlands below 7,500 feet along the coast from British Columbia to southern California. Shade to part shade, humusy soils, occasional to infrequent or no summer water. Deciduous and dormant in summer if grown dry. Other *Dryopteris* species need more water. USDA: 7–9 Sunset: N/A WUCOLS: L/M

^ *Dudleya caespitosa*

Dudleya

DUDLEYA

Succulents with rosettes of green to gray-green or gray-white leaves and clusters of small, cup-shaped or tubular flowers on tall stems. Some offset to form a small colony; others remain solitary. Best along the coast in cool sun with perfect drainage and infrequent to no summer water. Plant tilted to allow water to drain rapidly away. Good seaside plants. May be offered as *Echeveria*. USDA: 9–11 WUCOLS: L/VL

D. brittonii, giant chalk dudleya, to 1 foot tall and wide, with chalky, bluish white leaves and pale yellow flowers on reddish, 1- to 3-foot stems in summer. Does not form offsets. Native to steep, rocky cliffs and slopes along the coast of northwestern Baja California. Sunset: 16–17, 21–24

D. caespitosa, coast dudleya, 8–10 inches tall and 1 foot wide, with gray-white, oblong or sometimes fingerlike leaves with red tips and bright yellow flowers atop an 8-inch, red-tinted, branching stem in spring and summer. Spreads by offsets. Native to rocky cliffs in coastal California from Monterey to northern Los Angeles County. Sunset: 9, 14–17, 19–24

D. farinosa, bluff lettuce, 8–10 inches tall and 1 foot wide, with pale green or gray-green, broad and pointed leaves, often tinged pink or bright red, and clusters of pale to bright yellow flowers on reddish stems in spring. Spreads by offsets. Native to rocky

^ *Dudleya farinosa*

soils on coastal bluffs and hills in southwestern Oregon and northwestern California. Sunset: 5, 7, 14–17, 19–24

D. 'Frank Reinelt', 6–8 inches tall and 1 foot wide, with silvery blue-gray, fingerlike leaves that take on purplish tones in winter and yellow flowers in summer. Spreads by offsets. Hybrid involving *D. caespitosa*. May be offered as *D.* 'Anacapa'. Sunset: N/A

^ *Dymondia margaretae*

Dymondia margaretae

SILVER CARPET

Perennial, dense, tight mat, 1–2 inches tall and slowly spreading to 2–3 feet wide, with narrow, dark gray-green leaves, silvery beneath, and small, yellow daisy flowers in summer. Native to coastal plains of southwestern Western Cape Province, South Africa. Sun, good drainage, occasional to infrequent summer water. Best along the coast. USDA: 9–11 Sunset: 15–24 WUCOLS: L

^ *Echeveria agavoides*

^ *Echeveria elegans*

Echeveria

HEN AND CHICKS

Succulents with rosettes of green to gray-green leaves and nodding, bell-shaped flowers on tall stems in spring or early summer. Native to mid- to high-elevation, semi-arid parts of Mexico, Central America, and southwestern South America. Cool sun to part shade, well-drained, sandy or gravelly soils, occasional to infrequent summer water. Plant tilted so that water drains rapidly away. USDA: 9–11 WUCOLS: L

E. agavoides, 6 inches tall and 8–12 inches wide, spreading by offsets, with bright green leaves with red tips and a terminal spine and red and yellow flowers. 'Lipstick' has green leaves with red edges and tips. 'Ebony' has dark red-brown leaf edges and tips. Sunset: 8–9, 12–24

E. cante, 6–8 inches tall and 10–12 inches wide, with spoon-shaped, powdery blue-green leaves, often blushed pink and edged in red, and pink flowers with bright orange interiors on 18-inch stems. Does not form offsets. 'White Cloud' has especially white leaves. Sunset: N/A

E. elegans, hen and chicks, a few inches tall and wide, freely offsetting to form a mound to 8 inches tall and 1 foot wide, with silvery blue-green leaves and pink flowers with a yellow-orange tinge. Sunset: 8–9, 12–24

E. ×imbricata, 3 inches tall and 4–8 inches wide, with bluish gray-green leaves edged in pink and clusters of red and yellow flowers on tall stems. Offsets freely to form dense clumps. Hybrid of *E. secunda* and *E. gibbiflora* 'Metallica'. Plants with similar parentage are offered as 'Blue Rose', 'Compton Carousel', 'Gray Swirl', and 'Imbricata', or simply as blue rose echeveria. Sunset: 8–9, 12–24

Elaeagnus
SILVERBERRY

Evergreen shrubs with leathery, green leaves with tiny silvery spots and clusters of small, tubular, usually fragrant flowers in fall, followed by small, reddish brown fruit. Sun or shade, well-drained soils, occasional to infrequent or no summer water. *E. angustifolia* is invasive along the Pacific coast, especially in riparian areas. USDA: 7–9 Sunset: 4–24 WUCOLS: L/M

E. ×ebbingei*, see *E. ×submacrophylla

E. pungens, silverberry, 10–15 feet tall and wide, with dark olive-green, wavy-edged leaves on spiny branches and fragrant, creamy white flowers. Native to coastal areas of China and Japan, but adaptable inland. 'Fruitlandii' has a more symmetrical form and larger, more rounded, silvery leaves. 'Maculata' has creamy yellow leaves with green margins.

E. ×submacrophylla, oleaster, 8–12 feet tall and wide, with dark green leaves and fragrant, creamy white flowers. Hybrid between *E. macrophylla*, native to coastal areas of Korea and Japan, and *E. pungens*. 'Gilt Edge' has dark green leaves with golden yellow margins. May be offered as *E. ×ebbingei*.

^ *Elaeagnus ×submacrophylla* 'Gilt Edge'

^ *Elegia tectorum*

Elegia
CAPE RUSH

Perennials from rhizomes with stiff, narrow, grass-like stems, small brownish, papery bracts at nodes along the stems, and clusters of tiny, dark brown flowers at stem ends. Initially upright and then relaxing on all sides to create a graceful mound. New stems arise from the center. Native to Western Cape and Eastern Cape provinces, South Africa, near the coast and often in marshes or seeps, but thrive also in dryish soils. Two species are in the trade: *E. elephantina* is 4–6 feet tall and wide; *E. tectorum* is 3–4 feet tall and wide. *E. tectorum* 'El Campo' is a compact selection. Sun to light shade, most soils, occasional to infrequent summer water. May be offered as *Chondropetalum*. USDA: 9–10 Sunset: 8–9, 14–24 WUCOLS: L/M

Elymus
WILD RYE

Cool-season grasses that spread by rhizomes, some more aggressively than others. Widely distributed in temperate regions of Europe, Asia, and North America. Grown for their gray-green to blue-gray foliage. Sun to part shade, most soils, infrequent to no summer water. May be offered as *Leymus*.

E. cinereus, Great Basin or ashy wild rye, 3–6 feet tall and 2–4 feet wide, coarse-textured, with broad, stiffly erect, green or blue-green leaves and summer flowers on tall stems above the leaves. Native to dry, moist, or seasonally moist grasslands and woodland openings below 9,000 feet from British Columbia south to California, Arizona, and New Mexico and east to South Dakota. USDA: 5–9 Sunset: N/A WUCOLS: L

E. condensatus, giant wild rye, 3–6 feet tall and widely spreading, with bright green to bluish green leaves and tan flowers on stems well above the foliage in summer. Native to coastal sage scrub, chaparral, and foothill woodlands below 5,000 feet in coastal southern California and Baja California. 'Canyon Prince', a selection from the Channel Islands, is 2–3 feet tall with powdery blue-gray leaves. Cut to the ground in late winter to maintain upright form; new leaves are green and mature quickly to blue-gray. Floppy and fast spreading with summer water. USDA: 7–10 Sunset: 7–12, 14–24 WUCOLS: L

^ *Elymus condensatus*

Encelia
BRITTLEBUSH

Evergreen to semi-deciduous shrubs with oval to broadly lance-shaped, green to gray-green leaves and bright yellow daisy flowers with darker centers. Drop leaves in late summer if grown dry. Self-sow and may become invasive. Sun, well-drained soils, occasional to infrequent or no summer water. Cut back in late summer to renew. Need summer heat. WUCOLS: L/VL

E. californica, brittlebush or coast sunflower, 2–4 feet tall and wide, with medium green leaves and flowers from late winter to summer. Native to drier west- or south-facing slopes along the immediate coast and on the coastal side of inland foothills in southern California and northwestern Baja California. USDA: 9–10 Sunset: 7–16, 18–24

E. farinosa, brittlebush or incensio, 2–4 feet tall and 3–4 feet wide, with silvery greenish gray, woolly leaves and flowers in spring. Native to dry washes and sunny, dry slopes in the southwestern United States and northwestern Mexico. USDA: 8–10 Sunset: 8–16, 18–24

^ *Encelia californica*

Epilobium
CALIFORNIA FUCHSIA

Perennials with oval to lance-shaped, gray-green leaves and showy tubular flowers in mid- to late summer and fall. Spread by rhizomes and by seed. Deciduous or semi-deciduous in winter. Cool sun and infrequent to no summer water near the coast, afternoon shade and occasional water inland, most well-drained soils. May be offered as *Zauschneria*. WUCOLS: L/VL

E. 'Bowman's #1' ('Bowman's Best', 'Bowman's Hybrid', 'Bowman'), upright to 2–3 feet tall and wide, with narrow, olive-green leaves and masses of small, red-orange flowers. Hybrid of garden origin from northern California, likely between *E. septentrionale* and *E. canum*. USDA: 7b–11 Sunset: 5–7, 14–17, 19–24

^ *Epilobium canum*

E. canum, California fuchsia, prostrate to upright and 3–4 feet tall, with gray-green to silvery gray leaves and red-orange flowers. Native to a variety of coastal and inland habitats, from sea level to 10,000 feet, in California and Oregon east to Idaho and Wyoming and south to Arizona and New Mexico. 'Calistoga', a selection from Napa County in northern California, is 18 inches tall with especially large, velvety gray leaves. 'Catalina', from Catalina Island in southern California, is 3–4 feet tall with silvery gray leaves. 'Everett's Choice', a garden selection from northern California, likely of the variety *latifolium*, is less than 6 inches tall and 3–5 feet wide. USDA: 8–10 Sunset: 2–11, 14–24

E. septentrionale, Humboldt County fuchsia, low mats of silvery gray-green leaves and red-orange flowers. Native to rocky slopes and streambanks in northern California from sea level to 7,000 feet. 'Select Mattole', a selection from Humboldt County, is 6–8 inches tall and 2–3 feet wide, with broadly triangular, silvery gray leaves and large flowers. 'Wayne's Silver' is similar. Best near the coast. USDA: 8–11 Sunset: 5–7, 14–17, 19–24

Epimedium

EPIMEDIUM, BARRENWORT

Perennials from rhizomes, low and spreading slowly or vigorously, with leaves divided into heart-shaped, oval, to almost lance-shaped leaflets on short, thin, wiry stems. Sprays of small spring flowers on branched stems, barely above or just below the leaves. Cut back in late winter for a better view of flowers and new leaves. Part shade or shade, humusy, well-drained soils, occasional to infrequent summer water. Good for dryish shade. Sunset: 2–9, 14–17 WUCOLS: M

E. alpinum, less than 1 foot tall and spreading vigorously, with green leaves that emerge pink-tinged and turn reddish in fall and yellow and red flowers below and among the leaves. Native to woodlands in mountains of southern Europe. USDA: 5–9

E. grandiflorum, to 1 foot tall and slowly spreading, with medium green leaves, bronzy when new, and flowers in a range of colors above the leaves. Native to deciduous woodlands in China, Japan, and North Korea. Needs acidic soils and more summer water than southern European species. USDA: 5–9

E. ×***rubrum***, red barrenwort, to 1 foot tall and spreading vigorously, with green leaves edged with red, new growth and fall color bronzy red, and red and creamy yellow flowers above the leaves. Hybrid of garden origin between *E. alpinum* and *E. grandiflorum*. USDA: 4–9

^ *Epimedium* ×*rubrum*

Erica

HEATH

Evergreen shrubs, from low mats to mounding or tall, with fine-textured, needlelike leaves and small, lightly fragrant, tubular or cup-shaped flowers. Many heaths are particular about soils and many need regular summer water. Those listed here accept moderate to occasional summer water in cool sun or part shade and almost any well-drained soil. Not low water in hot-summer climates. Some spread by seed and can be weedy. WUCOLS: M

E. arborea, tree heath, 6–15 feet tall and 4–6 feet wide, with medium green leaves and small, cup-shaped, white to pale pink flowers in winter and spring. Native from the Canary Islands and Madeira through southern Europe and northern Africa to the Black Sea and south to Saudi Arabia and parts of east Africa. Plants in cultivation likely are from the Mediterranean region. 'Albert's Gold', 5 feet tall and 4 feet wide, has chartreuse new growth and white flowers. USDA: 7–9 Sunset: 15–17, 21–24

^ *Erica canaliculata*

E. canaliculata, Christmas heather, upright to 6–8 feet tall and 4–8 feet wide, with small, green to gray-green leaves and clusters of fragrant, pink flowers in fall and winter. Native to forest edges and valleys in coastal plains of southern Western and Eastern Cape provinces, South Africa. May be offered as *E. melanthera*. USDA: 9–10 Sunset: 15–17, 20–24

^ *Erica carnea*

E. carnea, winter or alpine heath, to 1 foot tall and 1–2 feet wide, with medium green leaves and white, pale pink, or lavender, tubular flowers in winter. Native to mountainous areas of central and southeastern Europe. 'Ruby Glow' has dark red flowers. One of the most cold-hardy heaths. USDA: 4–7 Sunset: 2–10, 14–24, A3

E. ×darleyensis, upright to 2–3 feet tall and wide, with medium green leaves and pink to lavender or white, cup-shaped flowers in winter and spring. Hybrid of garden origin, likely between *E. erigena* and *E. carnea*, discovered in England. Many cultivars in this adaptable group. USDA: 7–8 Sunset: 2–10, 14–24

Erigeron
FLEABANE

Annuals and perennials with pink, white, or lavender daisy flowers, usually with yellow centers. Bloom from spring to summer or almost year-round in mild climates. Those listed here are perennials. Full sun, most well-drained soils, infrequent summer water along the coast, part shade and moderate to occasional water inland.

E. divergens, spreading fleabane, less than 1 foot tall and 18 inches wide, with fine-textured, gray-green leaves and white flowers. Native to deserts and low mountains in much of western North America from southern British Columbia and Alberta south to Baja California and northern Mexico. USDA: 5–9 Sunset: N/A WUCOLS: M

E. glaucus, beach aster or seaside daisy, to 1 foot tall and 1–2 feet wide, with green to blue-green leaves and lavender to nearly white flowers. Native to coastal bluffs and dunes from Oregon to northern California. 'Bountiful', a selection from southern California, has especially large flowers. 'Cape Sebastian', a selection from the southern Oregon coast, is especially dense and compact. Best along the coast. USDA: 6–10 Sunset: 4–6, 15–17, 22–24 WUCOLS: L/M

^ *Erigeron glaucus*

E. ×moerheimii, pink Santa Barbara daisy, 1–2 feet tall and 2–3 feet wide, with small, narrow, green leaves and pink to pale lavender flowers. Similar to *E. karvinskianus* but does not reseed. Garden hybrid of uncertain origin, possibly involving *E. karvinskianus*; may be offered as *E. karvinskianus* 'Moerheimii'. USDA: 7b–11 Sunset: N/A WUCOLS: L/M

E. speciosus, aspen daisy, 2 feet tall and wide, with long, narrow, green leaves and lavender or purple flowers. Native to woodland and forest openings between 2,500 and 11,000 feet from southern British Columbia to northwestern Oregon east to South Dakota and south to Arizona and New Mexico. 'Darkest of All' has large, blue-violet flowers. USDA: 5–9 Sunset: 1–9, 14–24 WUCOLS: M

E. 'W.R.', 1 foot tall and 2 feet wide, with large, lavender flowers. Believed to be a hybrid involving *E. glaucus*. Good performance coastal and inland. May be offered as 'Wayne Roderick'. USDA: 7–10 Sunset: 4–6, 15–17, 22–24 WUCOLS: L/M

^ *Erigeron speciosus*

Eriogonum
BUCKWHEAT

Evergreen shrubs or shrubby perennials with green to gray-green or silvery gray leaves and domed or flat-topped clusters of tiny flowers in spring and summer. Flowers age to various shades of rusty red or brown, giving plants a multicolored effect in late summer and fall. Sun to part shade, good drainage, occasional to infrequent or no summer water.

E. arborescens, Santa Cruz Island buckwheat, 3–4 feet tall and 4–6 feet wide, with narrowly lance-shaped, woolly, pale green to gray-green leaves and creamy white to pale pink flowers. Native to sandy or rocky soils on the Channel Islands in southern California. USDA: 7–10 Sunset: 5, 7–9, 14–24 WUCOLS: L/VL

E. fasciculatum, California buckwheat, variable, 2–4 feet tall and 4–6 feet wide, with small, narrowly linear, leathery, green leaves and pink and creamy white flowers. Native to dry slopes and canyons in coastal central California, the southern Sierra Nevada, and southern California desert foothills east to Utah, Arizona, and northwestern Mexico. 'Bruce Dickinson', 'Warriner Lytle', and 'Theodore Payne' are low, widely spreading selections. USDA: 7–10 Sunset: 7–9, 12–24 WUCOLS: L/VL

E. giganteum, St. Catherine's lace, variable, 3–8 feet tall and wide, with branches sparsely covered in woolly, oval, silvery gray-green leaves and white or pinkish white flowers. Native to dry, rocky slopes below 1,500 feet on the Channel Islands in southern California. Adaptable to inland as well as coastal conditions. USDA: 8–10 Sunset: 5, 7–9, 14–24 WUCOLS: VL

E. latifolium, seaside buckwheat, 1 foot tall and 2 feet wide, with felted, oval, bluish white leaves and creamy white to pinkish white flowers. Native to dunes and bluffs along the coast from Washington to central California. Best along the coast. Part shade inland. USDA: 8–11 Sunset: N/A WUCOLS: L

E. umbellatum, sulfur buckwheat, variable, from a few inches to 3 feet tall and 1–3 feet wide, with woolly, oval, green to gray-green leaves, silvery gray beneath, and bright yellow to creamy white flowers on tall stems. Native to dry, open, usually rocky slopes from 1,200 to 10,000 feet and from southwestern British Columbia to southern California and east to Colorado, Arizona, and New Mexico. 'Shasta Sulfur', a selection from Mount Shasta in northern California, is 1 foot tall and 2 feet wide; it may be offered as a selection of subspecies *polyanthum*. Afternoon shade inland. USDA: 4–10 Sunset: 1–24 WUCOLS: L/M

^ *Eriogonum fasciculatum* 'Bruce Dickinson'

^ *Eriogonum giganteum*

Eriophyllum
WOOLLY SUNFLOWER

Perennials with green or gray-green leaves and golden yellow flowers on upright stems in spring and summer. Sun to part shade, good to excellent drainage, occasional or infrequent to no summer water. Cut back in late winter to renew. Sunset: N/A

E. confertiflorum, golden yarrow, 1–2 feet tall and 2–3 feet wide, variable plant with small, lobed or divided, green or silvery gray-green leaves, white-woolly beneath, and flowers in tight clusters. Native to dry, rocky soils in Sierra Nevada foothills, on coastal bluffs, and in coastal mountains from the San Francisco Bay Area to northern Baja California. Drops some leaves in summer if grown dry. USDA: 6–10 WUCOLS: L/VL

E. lanatum, woolly sunflower or Oregon sunshine, variable plant from less than 1 foot to 2 feet tall and 2–3 feet wide, with silvery gray-green, woolly, linear or lobed leaves and flowers singly on tall stems. Native to coastal bluffs and inland forest openings below 10,000 feet from British Columbia to central California and east to western Wyoming. Self-sows readily. Afternoon shade in hot locations. USDA: 5–10 WUCOLS: L

E. nevinii*, see *Constancea

E. staechadifolium, seaside woolly sunflower, 2–3 feet tall and 3–4 feet wide, with deeply lobed, gray-green leaves, silvery beneath, and flowers in tight clusters. Native to dunes and bluffs below 300 feet along the immediate coast from Santa Barbara County and the Channel Islands north to Oregon. May be short lived in southern California. USDA: 8–10 WUCOLS: L

< *Eriophyllum confertiflorum*

^ *Erodium reichardii*

Eriostemon, see *Philotheca*

Erodium
CRANESBILL

Perennials with low rosettes of green to gray-green, slightly scalloped leaves and clusters of small, pink or white flowers in spring and early summer. Many cranesbills need regular summer water. Those listed here thrive with moderate to occasional water in cool sun to part shade and with excellent drainage. Best in cool, coastal climates. Not low water in hot-summer climates. WUCOLS: L/M

E. chamaedryoides*, see *E. reichardii

E. chrysanthum, yellow cranesbill, mounding 6–8 inches tall and 1–2 feet wide, with silvery gray-green leaves and pale yellow flowers with silvery veins in spring and summer. Native to rocky, open sites from 4,500 to 7,000 feet in the mountains of Greece and Albania. USDA: 8–9 Sunset: N/A

E. reichardii, alpine geranium, 2–6 inches tall and 1 foot wide, with dark green to silvery green leaves and white flowers with purplish pink veins. Native to the Balearic Islands in the western Mediterranean. May be offered as *E. chamaedryoides*. Good rock garden plant. USDA: 6–9 Sunset: 7–9, 14–24

Eryngium
SEA HOLLY

Perennials and annuals with rosettes of basal leaves, sometimes deeply toothed and often spiny, and upright, leafy, branched stems topped by thistlelike, cylindrical cones of tiny summer flowers with usually spiny bracts. Plants listed here are perennials. Full sun, good to excellent drainage, moderate to occasional summer water.

E. alpinum, alpine sea holly, 1–2 feet tall and wide, with heart-shaped, spiny, bluish green leaves and steel blue flowerheads with bristly blue bracts. Native to rocky soils in the Alps between 5,000 to 6,500 feet from France to the northern Balkan Peninsula. Best where summer nights are cool. USDA: 5–9 Sunset: 2–24 WUCOLS: M

E. amethystinum, blue sea holly, 1–2 feet tall and wide, with deeply cut, spiny, gray-green leaves and blue flowerheads with long, spiny, silvery blue bracts. Native to rocky soils in the mountains of Italy and the Balkan Peninsula. USDA: 4–9 Sunset: 1–24 WUCOLS: M

E. planum, blue eryngo, 2–3 feet tall and wide, with coarsely toothed and rounded, dark green leaves and steel blue flowerheads on violet-blue stems with spiny, blue-green bracts. Native to open areas in gritty or sandy soils from central and southeastern Europe into Asia. USDA: 5–9 Sunset: 2–24 WUCOLS: L/M

E. 'Sapphire Blue', 2–3 feet tall and 18 inches wide, with intensely blue-violet flowers, stems, leaves, and bracts. Hybrid of unnamed cultivars of *E. bourgatii* and *E. alpinum* made in Holland. USDA: 5–9 Sunset: N/A WUCOLS: N/A

^ *Eryngium* 'Sapphire Blue'

Erysimum
WALLFLOWER

Annuals and short-lived perennials with narrow, lance-shaped leaves and often lightly fragrant flowers. Many wallflowers need regular summer water. Those listed here are perennials that thrive with moderate to occasional or infrequent summer water in sun to part shade and with excellent drainage.

E. capitatum, western wallflower, variable, 1–2 feet tall and wide, with grayish green leaves and golden yellow or orange flowers from spring to late summer or fall. Native to a wide range of habitats, from high mountains to low-elevation deserts, in much of western North America. Occasional to infrequent summer water. USDA: 7–10 Sunset: N/A WUCOLS: N/A

E. concinnum, Point Reyes wallflower, 1 foot tall and 18 inches wide, with dark green leaves and fragrant, creamy white to pale yellow flowers in spring. Native to coastal bluffs and rocky places from sea level to 1,000 feet from central California to southwestern Oregon. May be offered as a subspecies of *E. menziesii*, a native of coastal dunes in northern California. Occasional summer water. USDA: 8b–10 Sunset: N/A WUCOLS: L

E. franciscanum, San Francisco wallflower, 18 inches tall and wide, with dark green leaves and fragrant, creamy white to pale yellow flowers in late winter and spring. Native to the northern California coast from Sonoma County south to Santa Cruz County. Occasional summer water. USDA: 9–10 Sunset: N/A WUCOLS: N/A

E. linifolium, wallflower, 1–2 feet tall and wide, with lance-shaped, gray-green leaves and purple flowers in spring. Native to northwestern Spain and northern Portugal. 'Variegatum' has gray-green leaves with creamy white margins. Moderate summer water. USDA: 7–10 Sunset: 4–6, 14–24 WUCOLS: L/M

^ *Erysimum linifolium* 'Variegatum'

Eschscholzia californica

CALIFORNIA POPPY

Perennial, 1–2 feet tall and wide, with finely dissected, blue-green leaves and golden yellow to orange flowers in late winter to summer. Self-sows freely. Native to grassy, open areas from sea level to 6,500 feet from southern Washington to Baja California and east to New Mexico. The variety *maritima*, native along the coast, is more compact and has bright yellow flowers that shade to deep orange toward the center. Cultivars, available with white, red, or pink flowers, may not persist or reseed. Full sun, most soils, occasional to infrequent or no summer water. USDA: 8–10 Sunset: 1–24 WUCOLS: VL

^ *Eschscholzia californica*

Eucalyptus

EUCALYPTUS

Evergreen trees and large shrubs, fast growing, with aromatic, leathery, green to gray-green or blue-green leaves. Most have juvenile leaves that differ from mature leaves. All are native to Australia. Cold hardiness varies. The hardiest are native to the mountains of New South Wales and Tasmania in southeastern Australia. Full sun, most well-drained soils, infrequent to no summer water.

E. archeri, alpine cider gum, 25–40 feet tall and 15–20 feet wide, with gray-green, narrowly oval, lance-shaped leaves, juvenile leaves rounded and pale gray-blue. Grayish white, peeling bark and white spring flowers. Native to mountains of northeastern Tasmania between 3,000 and 4,700 feet. May be offered as a subspecies of *E. gunnii*, another high-elevation species from Tasmania. USDA: 8–10 Sunset: N/A WUCOLS: N/A

***E. citriodora*, see Corymbia**

E. forrestiana, fuchsia gum, 12–20 feet tall and 10–12 feet wide, with dark green, lance-shaped leaves, gray, peeling bark, and bright red and yellow flowers. Native to coastal plains in southern Western Australia. USDA: 9–11 Sunset: N/A WUCOLS: N/A

E. macrocarpa, mottlecah, sprawling shrub, 6–10 feet tall and wide, with silvery blue-gray, broadly oval leaves on gray-white branches. Bright red flowers in spring and summer followed by silvery seedpods. Native to sandy soils in southwestern Western Australia. USDA: 8–10 Sunset: N/A WUCOLS: N/A

***E.* 'Moon Lagoon'**, multistem shrub or small tree, 6–12 feet tall and 4–8 feet wide, with small, oval, silvery blue-green juvenile leaves, maturing to lance-shaped and olive-green, and white flowers in summer. Parentage uncertain, possibly a hybrid between *E. latens* and *E. kruseana*, two species from Western Australia. May be offered as a selection of *E. latens*. USDA: 9–11 Sunset: 5–6, 8–24 WUCOLS: L

E. nicholii, willowleaf peppermint, 35–50 feet tall and 15–30 feet wide, with especially narrow, lance-shaped, light green leaves on pendulous branches and thick, roughly furrowed, reddish brown bark that does not peel. White summer flowers. Native to grasslands and woodlands in the mountains of New South Wales. USDA: 8–10 Sunset: 5–6, 8–24 WUCOLS: L/M

E. pauciflora, snow gum, usually multitrunk, to 25–60 feet tall and 20–40 feet wide, with creamy white to pale gray, peeling bark, and silvery gray-green, broadly lance-shaped mature leaves. White flowers in spring and summer. Native to eastern and

^ *Eucalyptus macrocarpa*

^ *Eucalyptus* 'Moon Lagoon'

southeastern Australia. Two subspecies, *niphophila*, alpine snow gum, and *debeuzevillei*, Jounama snow gum, are found at the highest elevations in the mountains of New South Wales. USDA: 8–10 Sunset: N/A WUCOLS: N/A

E. perriniana, spinning gum, 20–30 feet tall and 15–30 feet wide, with round, blue-gray juvenile leaves that encircle the stem, bluish green, lance-shaped mature leaves, and smooth, peeling, off-white to gray-green bark. Small, white flowers in spring. Native to the mountains of New South Wales and Tasmania. USDA: 7–11 Sunset: N/A WUCOLS: N/A

Euphorbia

SPURGE

Evergreen or deciduous shrubs and perennials, many with succulent stems or leaves and some resembling cactus. Those listed here are herbaceous perennials with long, narrow, linear leaves and tiny, inconspicuous flowers surrounded by showy, colorful bracts in spring. Many perennial euphorbias self-sow prolifically or spread by rhizomes. A few, such as *E. esula*, *E. terracina*, and *E. oblongata*, are considered noxious weeds. Others, including *E. myrsinites*, *E. characias*, and *E. rigida*, are widely grown in gardens but are heavy self-seeders and easily escape into nearby wildlands. Cool sun to part shade or afternoon shade, good drainage, occasional to infrequent summer water. Milky white sap can irritate skin.

E. amygdaloides, wood spurge, upright to 2–3 feet tall and 1–2 feet wide, with dark green leaves, reddish purple beneath, and lime-green bracts. Native to northwestern Turkey, the Caucasus, and parts of the northwestern Mediterranean region. BLACKBIRD ('Nothowlee') has dark purple leaves. USDA: 6–9 Sunset: 2b–24 WUCOLS: M

^ *Euphorbia characias* subsp. *wulfenii*

E. characias, spurge, 3–4 feet tall and 2–3 feet wide, with blue-green leaves and chartreuse to yellow bracts. Native from Portugal and Spain to Turkey. Subspecies *wulfenii* has especially large flowerheads and bracts. 'Tasmanian Tiger' has gray-green leaves with creamy white edges. USDA: 7–10 Sunset: 4–24 WUCOLS: L/VL

^ *Euphorbia rigida*

E. rigida, 1–2 feet tall and 2–3 feet wide, with silvery blue-green leaves on upright stems and chartreuse bracts. Native to the Mediterranean region from Portugal to Italy, Greece, and Albania into Turkey. May be offered as *E. biglandulosa*. USDA: 7–10 Sunset: 4–24 WUCOLS: L

Euryops
SHRUB DAISY

Evergreen shrubs with deeply lobed, green or gray leaves and bright yellow daisy flowers at stem tips almost year-round. Sun to light shade, most soils, occasional to infrequent summer water. Cut back hard every few years or tip prune regularly to renew.

E. chrysanthemoides, African bush daisy, 3–5 feet tall and wide, with glossy, dark green leaves. Native along the coast and inland in Eastern Cape Province, South Africa. Similar to *E. pectinatus* 'Viridis'. Infrequent summer water. USDA: 8–11 Sunset: N/A WUCOLS: M

^ *Euryops pectinatus* 'Viridis'

E. pectinatus, euryops, 3–5 feet tall and wide, with gray-green leaves. Native to rocky slopes near the coast in Western Cape Province, South Africa. 'Munchkin' is 3 feet tall and wide. 'Viridis' is 4 feet tall and wide with green leaves. Occasional summer water. USDA: 9–11 Sunset: 8–9, 12–24 WUCOLS: L/M

Fallugia paradoxa
APACHE PLUME

Evergreen shrub, upright to 3–6 feet tall and 6 feet wide, with peeling whitish bark, small, deeply lobed, green leaves, and creamy white spring flowers. Showy clusters of fluffy pink seedheads with long, feathery plumes remain on the plant as petals fall. Native to dry, sandy, or rocky soils in desert mountains of southern California east to Texas and south to northern Mexico. Sun, well-drained soils, infrequent summer water. Drops leaves in summer if grown dry. USDA: 7–10 Sunset: 2–23 WUCOLS: VL/L/M

Feijoa, see *Acca*

Festuca
FESCUE

Cool-season bunchgrasses with narrow leaves and airy sprays of late spring to early summer flowers. Cool sun to part shade, good to excellent drainage, moderate to occasional or infrequent summer water. Best in cool-summer climates.

F. californica, California fescue, 2 feet tall and wide, with green to bluish green leaves and purple-tinged flowers on 4-foot stems. Native to open, brushy or wooded slopes, especially north-facing banks below 5,000 feet, in central and northwestern California and southwestern Oregon. 'River House Blues' may be especially well suited to southern California. USDA: 7–11 Sunset: 4–9, 14–24 WUCOLS: L/M

F. glauca, blue fescue, to 1 foot tall and wide, with fine-textured, blue-gray leaves and purple-tinged green flowers on arching stems above the foliage. Native to central and southern Europe. Fairly short lived. 'Elijah Blue' is less than 1 foot tall with silvery blue leaves; tends to turn brown in hot summer sun. May be offered as *F. cinerea* or *F. ovina* var. *glauca*. USDA: 4–8 Sunset: 1–24 WUCOLS: L/M

^ *Fallugia paradoxa*

^ *Fallugia paradoxa*

^ *Festuca californica*

F. idahoensis, Idaho fescue, to 1 foot tall and wide, with green to bluish green, fine-textured leaves and flowers on 2- to 3-foot stems. Native to open meadows and rocky slopes along the coast and in the mountains from British Columbia to central California and east to Montana, Idaho, and Colorado. 'Tomales Bay', from Marin County, California, is about 6 inches tall with deep blue-green leaves. USDA: 6-10 Sunset: 1-10, 14-24 WUCOLS: VL/L/M

F. 'Siskiyou Blue', 12-18 inches tall and 1-2 feet wide, with chalky, blue leaves and flowers that emerge green and mature to tawny gold. Garden origin, believed to be a hybrid between *F. idahoensis* and the European *F. glauca*. USDA: 4-9 Sunset: 1-10, 14-24 WUCOLS: L/M

^ *Forestiera pubescens*

Forestiera pubescens

DESERT OLIVE

Deciduous shrub or small tree, multitrunk, upright to 6-10 feet tall and 5-8 feet wide, with smooth, pale gray to almost white bark and lance-shaped to oval, light green leaves that turn yellow in fall. Clusters of tiny greenish yellow spring flowers and blue-black fruit. Native to rocky or sandy soils in washes and canyons from 300 to 5,500 feet and from southern California east to New Mexico and south to northern Mexico. Sun to light shade, good drainage, occasional summer water. Needs summer heat. May be offered as *F. neomexicana*. USDA: 6-10 Sunset: 1-3, 7-24 WUCOLS: L

^ *Fragaria vesca*

Fragaria

STRAWBERRY

Perennials with dark green, medium green, or bluish green leaves with reddish tints in winter, white flowers in late winter or spring, and red fruits in summer. Spread by surface runners, less often by seed. Sun to part shade, moderate to occasional summer water. WUCOLS: M

F. chiloensis, coastal strawberry, 4-8 inches tall, with glossy, dark green leaves and large fruit. Native to dunes and bluffs along the Pacific coast of North and South America. 'Aulon' has especially large leaves and flowers. Best in sandy or gritty soils. USDA: 7-10 Sunset: 4-24

F. vesca, woodland strawberry, 4-12 inches tall, with medium green leaves and small fruit. Native to woodland and forest habitats in much of temperate Europe, Asia, and North America. The subspecies *californica* is native to shady places below 6,000 feet in the Sierra Nevada and Coast Ranges in northern California and Oregon. Best in humusy soils. USDA: 5-9 Sunset: 2b-9, 14-24

< *Festuca idahoensis* (foreground)

^ *Frangula californica*

^ *Frangula californica* 'Leatherleaf'

F. virginiana* subsp. *platypetala, western alpine strawberry, low mat a few inches tall, with gray-green leaves that turn purplish in fall and small fruit. Native to the Sierra Nevada from 4,000 to 10,000 feet north to British Columbia. Best in rocky or fast-draining soils. USDA: 4–10 Sunset: N/A

Frangula californica
COFFEEBERRY

Evergreen shrub, 6–8 feet tall and wide, with dark green leaves, pale green beneath, insignificant greenish yellow flowers, and showy red fruits that age to black. Native to dry, sandy, or rocky slopes from southwestern Oregon to northern Baja California and east to western New Mexico. 'Eve Case', 'Mound San Bruno', and 'Leatherleaf' are all time-tested selections from coastal northern California. Sun to shade, most well-drained soils, occasional to infrequent or no summer water. May be offered as *Rhamnus*. USDA: 7–10 Sunset: 3–10, 14–24 WUCOLS: L/VL

Fremontodendron
FLANNEL BUSH

Evergreen shrubs with thick, leathery, felted and lobed, green leaves and masses of large, cup-shaped, golden yellow flowers in spring and summer. Tiny hairs on leaves and seedpods can be irritating to skin and eyes. Sun to part shade, fast drainage, infrequent to no summer water. USDA: 8–10 Sunset: 4–24 WUCOLS: VL

F. 'California Glory', 18–20 feet tall and 15–20 feet wide. Hybrid between *F. californicum* and *F. mexicanum*. *F.* 'San Gabriel' is similar.

F. californicum, flannel bush, fast growing to 15–20 feet tall and wide. Native to sandy washes and rocky slopes below 6,000 feet in coastal mountains and inland foothills in central to southern California and northern Baja California.

Fremontodendron 'California Glory' >

F. 'Dara's Gold', 3 feet tall and 6–8 feet wide. Hybrid between *F. decumbens* and *F. mexicanum*.

F. decumbens, Pine Hill or dwarf flannelbush, 2–3 feet tall and 6 feet wide. Native to Pine Hill in the Sierra Nevada foothills east of Sacramento. May be offered as a subspecies of *F. californicum*.

F. 'Ken Taylor', 4–5 feet tall and 6–8 feet wide. Hybrid between *F.* 'California Glory' and *F. decumbens*. *F.* 'El Dorado Gold' is similar.

F. mexicanum, Mexican flannel bush, 15–20 feet tall and 10–15 feet wide. Native to seasonal washes and canyon slopes from sea level to 3,000 feet in San Diego County and northwestern Baja California.

^ *Fremontodendron* 'Ken Taylor'

Furcraea

FURCRAEA

Succulent rosettes, with or without stems, of strap-shaped or sword-shaped leaves, with or without spines, and pendant flowers on tall stalks at maturity. Native to tropical regions of Mexico, the Caribbean, and northern South America. Most die after flowering, but numerous tiny plants that form on the flowering stalk fall to the ground and root. Sun to light shade, good drainage, infrequent summer water. USDA: 9b–11 Sunset: 13, 16–17, 19–24 WUCOLS: L

F. foetida, green aloe or Mauritius hemp, rosette 4–5 feet tall and 6–8 feet wide, stemless or short-stemmed, with long, slightly wavy, strap-shaped, spineless leaves and fragrant, greenish white flowers on a 15- to 20-foot stalk. Native to the Caribbean and northern Brazil. 'Mediopicta' has green leaves with a broad, creamy white stripe down the center. May be offered as *F. gigantea*.

^ *Furcraea macdougallii*

F. macdougallii, Macdougall's century plant, rosette 6–8 feet tall and wide, with narrow, upright, gray-green leaves, slowly forming an unbranched stem 8–12 feet tall, and greenish white flowers on a stalk rising 15–20 feet. Native to dry forests in the mountains of south-central Mexico. May be offered as *F. macdougalii*.

Galvezia, see Gambelia

Gambelia speciosa

ISLAND SNAPDRAGON

Evergreen shrub, mound of arching stems to 3–4 feet tall and 5–6 feet wide, with soft, slightly succulent, pale green leaves and bright red, tubular or trumpet-shaped flowers from spring to early fall. Native to bluffs and canyons below 3,000 feet on the Channel Islands in southern California and Guadalupe Island, off the coast of Baja California. 'Firecracker' is a compact selection. Sun to part shade or afternoon shade, most well-drained soils, infrequent to no summer water near the coast, occasional water inland. Cut back in late winter to renew. May be offered as *Galvezia*. USDA: 10–11 Sunset: 14–24 WUCOLS: L/VL

^ *Gambelia speciosa*

Garrya

SILKTASSEL

Evergreen shrubs with leathery, dark green leaves and long, pendant catkins in winter or early spring. Cool sun to part shade, most well-drained soils, infrequent to no summer water in cool-summer climates, afternoon shade and occasional water inland.

G. elliptica, coast silktassel, 10–15 feet tall and wide, with dark green, wavy-edged leaves and silvery green catkins. Native to dry, shaded or open slopes from sea level to about 1,500 feet along the coast from southern Oregon to northern Los Angeles County. 'James Roof' has foot-long catkins. 'Evie' is a denser shrub, 8–12 feet tall and wide. USDA: 7–10 Sunset: 4–9, 14–24 WUCOLS: L/M

^ *Garrya elliptica*

G. fremontii, Fremont's silktassel, 6–8 feet tall and wide, with smaller leaves than *G. elliptica* and shorter, silvery green catkins. Native to mountains and foothills along the coast as well as inland, from the Columbia River in southern Washington to parts of southern California. Good choice for hot inland areas as well as colder climates. USDA: 6–9 Sunset: 3–10, 12, 14–17 WUCOLS: L/VL

^ *Garrya elliptica* 'Evie'

Geranium ×cantabrigiense 'Biokovo'

Geranium macrorrhizum

G. ×issaquahensis, 6–12 feet tall and wide, with catkins with alternating bands of silvery green and reddish pink. Hybrid between *G. elliptica* and *G. fremontii*. 'Pat Ballard', a garden selection from northwestern Washington, and 'Glasnevin Wine', selected in Ireland, may be slightly more cold hardy than *G. elliptica*. USDA: 7–9 Sunset: N/A WUCOLS: N/A

Geranium

CRANESBILL

Perennials with lobed or finely divided, gray or green leaves and pink, purple, or white flowers in spring and summer. Spread slowly or vigorously by rhizomes and sometimes also by seed. Many geraniums need regular water. Those listed here thrive with moderate to occasional or infrequent water with good drainage in cool sun to part shade. WUCOLS: L/M

G. ×cantabrigiense 'Biokovo', 6–10 inches tall and slowly spreading, with dark green leaves and white flowers tinged pink at the base of each petal. Selection of a natural hybrid between *G. macrorrhizum* and *G. dalmaticum*, both native to mountains of the Balkan Peninsula. Occasional summer water. USDA: 4–9 Sunset: 1–24

G. incanum, carpet geranium, less than 1 foot tall and fast spreading, with finely cut, green leaves and small, lavender flowers. Native to South Africa. Self-sows readily and can be weedy. Best along the coast with infrequent summer water. May be short lived inland. USDA: 9–11 Sunset: 14–24

G. macrorrhizum, bigroot geranium, to 1 foot tall and spreading by rhizomes and by seed, with gray-green leaves and purplish pink or white flowers. Native to the Alps and the Balkans. Best in light shade with occasional summer water. USDA: 3–8 Sunset: 1–24

G. ROZANNE **('Gerwat')**, 1–2 feet tall, with marbled green leaves and large, blue-violet flowers with white centers. Hybrid of garden origin likely involving *G. himalayense* and *G. wallichianum* 'Buxton's Variety'. Spreads by rhizomes but does not self-sow. Needs moderate summer water. USDA: 5–8 Sunset: 2–11, 14–24

^ *Gilia capitata*

Gilia

GILIA

Annuals with finely divided, dark green leaves and fragrant flowers in spring and summer. Sun to part shade, well-drained soils, infrequent to no summer water. Self-sow. Need moisture in spring to bloom well. USDA: N/A Sunset: 1–24 WUCOLS: N/A

G. capitata, globe gilia, 1 foot tall and wide, with spherical clusters of tiny bright blue to lavender-blue flowers. Native to coastal dunes or mountain slopes from Alaska to northern Mexico and east to parts of Idaho, Arizona, and New Mexico.

G. tricolor, bird's eye gilia, 1–2 feet tall and 1 foot wide, with pale blue-violet and white, tubular to widely bell-shaped flowers with a ring of dark purple in the center. Native to sandy or rocky soils in California's Central Valley, Sierra Nevada foothills, and coastal mountains, often in seasonally moist meadows.

^ *Gilia tricolor*

Glandularia

MOCK VERBENA

Annuals and perennials with finely dissected, green leaves and clusters of lightly fragrant, star-shaped flowers. Those listed here are perennials. May be offered as *Verbena*.

G. bonariensis*, see *Verbena

^ *Glandularia lilacina* 'De la Mina'

G. gooddingii, Mojave verbena, 18 inches tall and 2–3 feet wide, with small, rounded clusters of tiny pinkish lavender flowers in late winter or spring. Self-sows. Native to sandy or rocky soils between 4,000 and 6,000 feet in deserts of the southwestern United States and northern Mexico. Sun to part shade, fast drainage, occasional to infrequent summer water. USDA: 6–9 Sunset: 7–24 WUCOLS: L/VL

G. lilacina, lilac verbena, mounding to 2–3 feet tall and 3–4 feet wide, with lavender flowers in spring and summer. Native to sandy or rocky soils on mountain slopes and in canyons in Baja California and on adjacent Cedros Island. 'De la Mina', Cedros Island verbena, has darker lavender flowers. Cool sun to light shade, good drainage, infrequent summer water. USDA: 7–10 Sunset: 12–24 WUCOLS: L

G. rigida*, see *Verbena

Glaucium

HORNED POPPY

Perennials, low rosettes of deeply incised, wavy-edged leaves and cup-shaped flowers on tall stems, followed by long, curved seedpods. Self-sow freely and can be weedy. Full sun, well-drained soils, occasional summer water. Good seaside plants. Sunset: N/A WUCOLS: L/VL

G. flavum, yellow horned poppy, 1–2 feet tall and wide, with silvery gray-green to blue-gray leaves and silky, golden yellow flowers on branched stems in spring and summer. Native to coastal regions of northern and southern Europe, northern Africa, and western Asia. The variety *fulvum* has silvery white leaves and bright orange flowers. USDA: 6–10

^ *Glaucium grandiflorum*

^ *Graptopetalum paraguayense*

G. grandiflorum, horned poppy, variable, 1–2 feet tall and wide, with silvery blue-green leaves and bright orange flowers, with a dark blotch at the base of each petal, in summer and fall. Native to open woodlands and shrublands of the eastern Mediterranean region from Turkey to Israel and east to Iran. USDA: 7–10

Graptopetalum paraguayense
GHOST PLANT

Succulent rosette, 4–5 inches tall and wide, with grayish white to bluish gray leaves, new growth tinged pink, and small yellow flowers in summer. Offsets form on gradually elongating stems. Origin uncertain; believed to be native to semi-arid, east-central Mexico. Sun or part shade, well-drained soil, occasional to infrequent or no summer water. Afternoon shade inland. USDA: 8–11 Sunset: 17, 19–24 WUCOLS: L/VL

^ *Grevillea* 'Austraflora Fanfare'

Grevillea
GREVILLEA

Evergreen shrubs with a wide range of shapes, sizes, and colors of leaves and flowers. Most are native to Australia, where they grow in many different habitats and climates, from desert to rainforest and from seaside to high mountains. Most flower in winter and spring. Sun to light shade, fast-draining soils, occasional to infrequent or no summer water. WUCOLS: L

G. 'Austraflora Fanfare' ('Fanfare'), less than 1 foot tall and 10–12 feet wide, with large, deeply cut, dark green leaves, bronzy when new, and dark red flowers. Hybrid of garden origin, believed to be a cross between *G.* ×*gaudichaudii* and *G. longifolia* (*G. aspleniifolia* var. *longifolia*), a species native to moist areas in forests of coastal New South Wales. USDA: 8b–10 Sunset: 15–24

G. australis, alpine grevillea, variable, from prostrate to 4–5 feet tall and wide, with small, narrow, olive-green leaves, silvery gray beneath, and fragrant, creamy white flowers. Native to mountains of southeastern Australia from Queensland to Tasmania. USDA: 7–10 Sunset: 15–16, 23–24

G. ×***gaudichaudii***, less than 1 foot tall and 10–12 feet wide, with dark green, deeply lobed, almost oaklike leaves, new growth dark red, and purplish red flowers. Natural hybrid between *G. laurifolia* and *G. acanthifolia*, both from the mountains of southeastern Australia. Best with some shade inland. USDA: 8–11 Sunset: N/A

G. juniperina, juniper grevillea, prostrate to upright and 6 feet tall, with bright green, prickly, needlelike leaves and flowers in a wide range of colors. Native to woodland and open forest at mid- to high elevations in the mountains of southeastern Australia. 'Low Red' ('Lava Cascade') is 1–2 feet tall and 5–8 feet wide, with red-orange flowers; drapes over walls. 'Molonglo' is similar, with yellow flowers. USDA: 8–10 Sunset: N/A

G. 'Kings Fire', 5–6 feet tall and 6–8 feet wide, with dissected, gray-green leaves on white stems and bright red flowers maturing to red-orange. Hybrid involving *G. nivea* ('Scarlet King') from coastal southern Western Australia, and 'Crowning Glory' ('Lasseter's Gold'), itself a hybrid of complicated parentage. USDA: 9–11 Sunset: N/A

G. lanigera, woolly grevillea, some forms prostrate and others mounding, to 4–6 feet tall and 6–10 feet wide, with narrow, soft, gray-green leaves and pink and white flowers. Native to southeastern Australia, both along the coast and in the mountains. 'Coastal Gem' is 1 foot tall and 4–5 feet wide. USDA: 9–10 Sunset: 15–24

G. lavandulacea, lavender grevillea, variable plant with softly felted, needlelike, silvery gray-green leaves and pink, red, or white flowers. Native to the mountains of southern Australia from South Australia to Victoria. 'Billywing' ('Billy Wings'), 1–3 feet tall and 3–4 feet wide, has bright red flowers; needs perfect drainage. 'Penola', 4–5 feet tall and 6–8 feet wide with reddish pink flowers, may be a hybrid with *G. lanigera*. USDA: 9–11 Sunset: 15–24

G. 'Neil Bell', 6–8 feet tall and wide, with lance-shaped, green leaves and large, pendant, red flowers. Willamette Valley chance seedling of *G.* 'Constance', a hybrid between *G. victoriae* and *G. juniperina*. Good choice for the Pacific Northwest. USDA: 8–10 Sunset: N/A

^ *Grevillea juniperina* 'Molonglo'

^ *Grevillea lavandulacea* 'Penola'

Hakea

HAKEA

Evergreen shrubs or small trees with many different shapes and sizes of leaves, flowers, and seedpods. Sun, well-drained soils, infrequent to no summer water. Good coastal plants. USDA: 9–11 Sunset: 9, 12–17, 19–24

H. francisiana, grassleaf hakea, 10–15 feet tall and 6–12 feet wide, with long, broad, linear, silvery green leaves on upright branches and coral-pink winter flowers in clusters resembling bottle brushes. Native to sandy or gravelly soils in arid and semi-arid regions of southern Western Australia and western South Australia. Does not accept heavy pruning. May be offered as *H. multilineata*. WUCOLS: N/A

H. laurina, pincushion hakea, 12–20 feet tall and wide, with lance-shaped, sharply pointed, bluish green leaves resembling eucalyptus and pale pink or white flowers in spherical heads in late fall to midwinter. Native to sandy soils along the south coast of Western Australia. WUCOLS: L

H. petiolaris, sea urchin hakea, 8–18 feet tall and 6–12 feet wide, with oval, sharply pointed, pale gray-green leaves and spherical clusters of pink and creamy white flowers in late fall or early winter. Native to southwestern Western Australia. WUCOLS: N/A

H. suaveolens, sweet hakea, 12–20 feet tall and 8–12 feet wide, with stiff, pale green leaves divided into sharply pointed, needlelike leaflets resembling a conifer and clusters of small, fragrant, white flowers in winter. Native along the south coast of Western Australia. Can be weedy. Potentially invasive in coastal areas. May be offered as *H. drupacea*. WUCOLS: L

Hakea laurina

Halimium atriplicifolium

Halimium

YELLOW ROCK ROSE

Evergreen shrubs with dark green to silvery gray-green leaves and yellow or rarely white flowers in spring or summer. Flowers drop their petals in late afternoon, replaced by new flowers the next morning. Native to the western Mediterranean region. Full sun, fast-draining soils, occasional to infrequent summer water. Drop some leaves in summer if grown dry. Sunset: N/A WUCOLS: L

H. atriplicifolium, 4–6 feet tall and 3–5 feet wide, with broadly oval, silvery gray-white leaves. Native from central and southern Spain to Morocco. USDA: 9–10

H. lasianthum, woolly rock rose, 1–3 feet tall and 2–3 feet wide, with felted, silvery gray-green leaves and flowers with a maroon or dark brown blotch at the base of each petal. Native to gritty soils in southern Portugal and southern Spain. *H. lasianthum* subsp. *formosum* 'Sandling' is a low-growing selection. USDA: 7–10

H. ×pauanum, 4–6 feet tall and 4 feet wide, with silvery green leaves and especially large flowers over a long season. Hybrid between *H. lasianthum* and *H. halimifolium*, a species native from Portugal to Italy and Morocco to Algeria. USDA: 8–10

^ *Hardenbergia violacea* 'Happy Wanderer'

^ *Hardenbergia violacea* 'Mini Haha'

Hardenbergia

LILAC VINE

Evergreen vines, fast growing to 10–15 feet, with dark green leaves and pendant clusters of lavender-purple flowers in midwinter to spring, followed by gray-brown seedpods. Climb by twining stems. Sun along the coast, part shade inland, most well-drained soils, occasional to infrequent summer water. USDA: 9–11 WUCOLS: M/L

H. comptoniana, with trifoliate leaves, giving the plant a fine-textured appearance. Native to sandy soils in southern and western Western Australia. Best in part shade. Sunset: 15–24

H. violacea, with lance-shaped leaves, slightly squared at the tip. Native to eastern and southeastern Australia from the coast into the mountains. 'Happy Wanderer' has larger leaves and flowers. 'Canoelands' has an especially long flowering period, extending into early summer. 'Mini Haha' is more shrublike, 2–3 feet tall and 3–5 feet wide. Sunset: 8–24

Hebe

HEBE

Evergreen shrubs with rounded to lance-shaped or scalelike leaves and clusters of tiny flowers in spring or summer. Most are native to a range of usually open and exposed New Zealand habitats from sea level to alpine or subalpine regions. Many are frost tender, and many need regular summer water. All need excellent drainage and good air circulation. Listed here are a few cold-hardy selections that do well in cool sun to light shade with moderate to occasional summer water. Best along the coast. May be offered as *Veronica*. Sunset: N/A WUCOLS: M

***H.* 'Emerald Gem'**, less than 1 foot tall and wide, with tiny, yellow-green, scalelike leaves and white flowers that appear only rarely. Rounded and tightly compact form opens up over time. Likely the same plant as *H.* 'Emerald Green' and *H.* 'Green Globe'. Natural hybrid discovered in the mountains of North Island, possibly involving *H. odora*, with tiny, spear-shaped leaves, and *H. subsimilis*, with scalelike leaves. USDA: 8–10

***H.* 'Karo Golden Esk'**, 2 feet tall and wide, with upright stems bearing green, scalelike leaves that turn golden in winter and tiny white flowers. Natural hybrid between *H. odora*, with tiny, spear-shaped leaves, and *H. armstrongii*, with scalelike leaves, discovered on South Island near the Esk River. Karo is an acronym for "known and recorded origin." USDA: 7–10

H. pimeleoides, 1–2 feet tall and 2–3 feet wide, with lance-shaped, silvery gray to bluish gray leaves and pale lilac flowers. Native to dry, open mountain grasslands on South Island. Usually available as 'Quicksilver', with lavender flowers. 'Western Hills', a garden selection from northern California, has white flowers. USDA: 8–10

^ *Hebe* 'Karo Golden Esk'

Helianthemum nummularium
SUNROSE

Evergreen shrub, 8–10 inches tall and 2–3 feet wide, with small, green or gray-green, lance-shaped leaves and masses of flowers in spring and early summer. Native to Europe and Asia Minor. Usually available as cultivars. 'Henfield Brilliant' has dark, silvery green leaves and bright orange flowers. 'Wisley Primrose' has gray-green leaves and bright yellow flowers. 'The Bride' has dark gray-green leaves and white flowers. 'Ben Hope' has gray-green leaves and rose-pink flowers. Full sun, well-drained soils, moderate to occasional summer water. Good seaside plant. USDA: 7–10 Sunset: 2b–9, 14–24 WUCOLS: L/M

^ *Helianthemum nummularium* 'Ben Hope'

Helichrysum, see *Ozothamnus*

Helictotrichon sempervirens
BLUE OAT GRASS

Cool-season bunchgrass, 2 feet tall and 2–3 feet wide, symmetrical arrangement of blue-gray leaves and bluish green midsummer flowers on 3- to 4-foot stems aging to tan. Native to southwestern Europe, from France to Italy. Cool sun to part shade, excellent drainage, occasional to infrequent summer water. Best in cool-summer climates. May be short lived, especially in heavy soils. USDA: 4–8 Sunset: 1–12, 14–24 WUCOLS: L/M

^ *Helictotrichon sempervirens*

Hesperaloe parviflora
RED YUCCA

Succulent, 3–4 feet tall and 4–6 feet wide, rosette of stiff, leathery, grasslike leaves with no spines or teeth and with white, fraying fibers on leaf margins. Rose-pink tubular flowers on arching, 5- to 6-foot stalks in summer. Native to rocky slopes and flats in deserts of southeastern Texas and northeastern Mexico. BRAKELIGHTS ('Perpa') is 2 feet tall with dark red flowers. Sun to light shade, well-drained soils, occasional to infrequent summer water. USDA: 6–10 Sunset: 2b–3, 7–16, 18–24 WUCOLS: L/VL

^ *Hesperaloe parviflora* BRAKELIGHTS

Hesperocyparis
CYPRESS

Evergreen coniferous trees or large shrubs with green to gray-green or blue-green, aromatic, scalelike leaves and small, usually round cones. Full sun, most well-drained soils, infrequent to no summer water.

H. arizonica, Arizona cypress, upright to 40–50 feet tall and 15–20 feet wide, with gray-green to blue-green leaves and reddish brown cones. Native to dry, rocky mountain slopes below 3,000 feet from southern California to western Texas and northern Mexico. *H. arizonica* var. *glabra* 'Blue Ice' has silvery blue-white leaves. May be offered as *Callitropsis* or as *Cupressus*. USDA: 7–10 Sunset: 7–24 WUCOLS: L/VL

H. forbesii, tecate cypress, upright and symmetrical, 20–30 feet tall and 15–20 feet wide, with light green leaves and colorful peeling bark. Native to dry slopes from 1,500 to 5,000 feet in coastal mountains of southern California and Baja California. Thrives inland as well as near the coast. May be offered as a variety of *C. guadalupensis*, as *Callitropsis*, or as *Cupressus*. USDA: 6–10 Sunset: 8–14, 18–20 WUCOLS: L/VL

^ *Hesperocyparis arizonica* var. *glabra* 'Blue Ice'

H. sargentii, Sargent's cypress, fast growing to 30–50 feet tall and 10–15 feet wide, with gray-green leaves and gray bark. Native to coniferous forest and chaparral in the Coast Ranges below 4,000 feet from Mendocino County to Santa Barbara County, California. May be offered as *Callitropsis* or as *Cupressus*. USDA: 7–10 Sunset: N/A WUCOLS: L

^ *Hesperoyucca whipplei*

Hesperoyucca whipplei
CHAPARRAL YUCCA

Succulent, 3–4 feet tall and 3–6 feet wide, stemless rosette of narrow, rigid, gray-green leaves with a sharp terminal spine and bell-shaped, creamy white to purplish flowers on a 10- to 15-foot stalk after about ten years. Dies after flowering but may be replaced by offsets. Native to dry, rocky soils from coastal southern California to northwestern Baja California and east to the western slopes of the southern Sierra Nevada from 1,000 to 8,000 feet. Not easy in cultivation outside its native range. Full sun, excellent drainage, infrequent to no summer water. May be offered as *Yucca*. USDA: 8–11 Sunset: 2–24 WUCOLS: L/VL

Heteromeles arbutifolia
TOYON

Evergreen shrub or small tree, usually multitrunk, 8–15 feet tall and 6–10 feet wide, with leathery, dark green leaves, clusters of creamy white flowers at branch ends in summer, and bright red berries in winter. Native to woodlands and chaparral below 4,000 feet in mountains of coastal California and Sierra Nevada foothills north to southern Oregon and south to Baja California. 'Davis Gold', grown in Davis, California, has bright yellow berries. Sun along the coast to part shade inland, most well-drained soils, good air circulation, infrequent to no summer water. USDA: 8–10 Sunset: 5–9, 14–24 WUCOLS: L/VL

^ *Heteromeles arbutifolia*

Heuchera

ALUMROOT, CORAL BELLS

Perennials from rhizomes, with rosettes of heart-shaped or rounded, lobed or scalloped leaves and airy clusters of tiny, bell-shaped flowers on tall stems in spring and summer. Many heucheras need regular summer water. Those listed here thrive with moderate to occasional or infrequent water in cool sun to shade and well-drained soils.

H. elegans, urn-flower alumroot, less than 1 foot tall and 1–2 feet wide with bright green leaves and drooping clusters of white flowers with pink bracts on 8- to 10-inch stems. Native to partly shaded, rocky slopes in the San Gabriel Mountains northeast of Los Angeles. May be offered as *H. caespitosa*. USDA: 8–10 Sunset: N/A WUCOLS: L/M

H. maxima, island alumroot, 1–2 feet tall and 2–3 feet wide, with large, bright green leaves, often mottled with gray or pale green, and pinkish white or creamy white flowers on maroon, 3-foot stems. Native to the northern Channel Islands in southern California. Best along the coast. USDA: 8–10 Sunset: 15–24 WUCOLS: L/M

H. sanguinea, coral bells, 1–2 feet tall and wide, with mottled green leaves and showy, deep pink to red flowers on 1- to 2-foot stems. Native to mountains of New Mexico and Arizona south into northern Mexico. 'Cinnabar Silver' has silvery green leaves with purple veins and bright red flowers. Moderate summer water. USDA: 4–10 Sunset: 1–11, 14–24, A1–3 WUCOLS: M

Hybrids of *H. maxima* and *H. sanguinea* display the tall stems of the former and the larger, darker colored flowers of the latter. Some hybrids prefer regular water. Among those that need less are *H.* 'Lillian's Pink', a garden hybrid from San Francisco with large, pink flowers; *H.* 'Rosada', with pink and creamy white flowers; *H.* 'Santa Ana Cardinal', with red flowers; and *H.* 'Wendy', with soft pink flowers. USDA: 8–10 Sunset: 14–24 WUCOLS: M/H

^ *Heuchera* 'Wendy'

^ *Heuchera* 'Rosada'

Hybrids involving *H. elegans* and *H. sanguinea* have the diminutive size of the former and the showy flowers of the latter. Hybrids in the Canyon Quartet series are 3–6 inches tall and 1–2 feet wide. Some prefer regular summer water. Among those that thrive with less are *H.* 'Canyon Melody', with white flowers, and *H.* 'Canyon Pink', with rose pink flowers. USDA: 7–10 Sunset: 2–11, 14–24 WUCOLS: M/H

Holodiscus discolor

OCEANSPRAY

Deciduous shrub, fast growing to 4–10 feet tall and wide, with triangular, deep green leaves, hairy beneath, and long sprays of tiny, fragrant, creamy white flowers from pink-tinged buds in late summer. Native to forests and shrublands in coastal mountains from southern British Columbia to southern California and east to Montana and the northern Rocky Mountains. Light shade, most soils, infrequent or no summer water. USDA: 5–11 Sunset: 1–9, 14–19 WUCOLS: L/M

^ *Holodiscus discolor*

Hunnemannia fumariifolia

MEXICAN TULIP POPPY

Perennial, to 1–2 feet tall and wide, with finely divided, blue-green leaves and bright yellow flowers with crinkled petals from midsummer through fall. Self-sows lightly. Native to the high deserts of northern Mexico, southwestern Texas, and Arizona. Full sun, most well-drained soils, occasional to infrequent or no summer water. USDA: 8b–11 Sunset: 1–24 WUCOLS: L/M

^ *Hunnemannia fumariifolia*

Hylotelephium, see *Sedum*

Iris

PACIFIC COAST IRIS

Perennials from rhizomes with sword-shaped to narrow, grasslike leaves and colorful, intricately patterned, sometimes lightly fragrant flowers in late winter or early spring. Dormant but often not deciduous in summer. Those listed here are native to the Pacific coast of North America. Cool sun to part shade, afternoon shade inland, well-drained soil, occasional to infrequent or no summer water.

^ *Iris* 'Canyon Snow'

^ *Iris douglasiana*

Cultivars may need more summer water than species. Sunset: 4–9, 14–24 WUCOLS: L/M

***I.* 'Canyon Snow'**, with dark green leaves and bright white flowers with golden yellow markings. Direct seedling of *I. douglasiana*. Large plant with large flowers, especially accepting of garden conditions. USDA: 7–11

I. douglasiana, Douglas iris, with strap-shaped, glossy green leaves and light blue-violet to dark purple or sometimes white flowers with purple, blue, or gold veining. Native to low-elevation coastal bluffs and grasslands from central western Oregon to southern California. Hybridizes readily in the wild. Likely the species most accepting of high winter rainfall and heavy soils. USDA: 5–10

I. innominata, Del Norte iris or golden iris, with narrow, grasslike leaves and bright golden yellow or creamy yellow flowers. Native to the Klamath Mountains in southwestern Oregon and northwestern California. Purple-flowered plants offered as *I. innominata* may be hybrids with *I. douglasiana*. USDA: 7–10

I. tenax, Oregon iris, dense clumps of narrow, fibrous, light green leaves and blue, purple, or sometimes white flowers on stems shorter than leaves. Native to grasslands and open woodlands in southwestern Washington and Oregon west of the Cascades. Likely the most cold-hardy species, but needs especially fast drainage. USDA: 7–10

^ *Jasminum nudiflorum*

Jasminum
JASMINE

Evergreen and deciduous shrubs, vines, and vining shrubs with white or yellow, sometimes fragrant flowers. Some jasmines prefer regular water. Those listed here thrive with moderate to occasional or infrequent summer water in sun to part shade and with good drainage.

J. mesnyi, primrose jasmine, evergreen vining shrub, 6–10 feet tall and wide, with dark green leaves on arching branches and sometimes fragrant, yellow flowers from early spring into summer. Native to southwestern China. Cut back hard occasionally to renew. May be offered as *J. primulinum*. Occasional summer water. USDA: 8–11 Sunset: 4–24 WUCOLS: L/M

J. nudiflorum, winter jasmine, deciduous vining shrub, 4–6 feet tall and 8 feet wide, with glossy, green leaves on arching branches and unscented, bright yellow flowers in winter. Needs support to climb. Cascades over walls. Native to northern China. 'Mystique' is 2 feet tall and 3 feet wide, with green leaves edged in white. Infrequent summer water. USDA: 6–10 Sunset: 2–21 WUCOLS: L

J. officinale, Spanish or poet's jasmine, deciduous or semi-evergreen vine, to 15–20 feet, with medium green leaves and fragrant white flowers from midspring through fall. Native to the Himalayas and the Caucasus. May be offered as *J. officinale* var. *grandiflorum* or as *J. grandiflorum*. Plants offered as *J. officinale* f. *affine* or as *J.* 'Affine' have large flowers with broader petals. Moderate summer water. USDA: 7–10 Sunset: 5–9, 12–24 WUCOLS: L/M

^ *Jubaea chilensis*

^ *Juncus patens*

Jubaea chilensis

CHILEAN WINE PALM

Palm, slow growing to 50–80 feet tall and 20–25 feet wide, with featherlike, gray-green leaves, 8–12 feet long, and a broad, thick trunk. Small purple flowers on a 4-foot stalk followed by round fruit, green when new and ripening to bright yellow. Native to dry woodlands in the mountains of central Chile. One of the hardiest palms, accepting of wet winters, and fine along the coast in Oregon. Old leaves drop cleanly on mature trees. Sun to part shade, well-drained soil, infrequent summer water. USDA: 8b–10 Sunset: 12–24 WUCOLS: L/M

Juncus patens

CALIFORNIA GRAY RUSH

Evergreen grasslike plant, to 2 feet tall and wide, with stiffly upright, gray-green to bluish green, leafless stems and tiny, rusty brown flowers partially hidden among the stems. Native to moist or seasonally moist locations near the coast below 5,000 feet from southern Washington to northern Baja California. 'Elk Blue', from coastal northern California, has especially narrow and especially blue stems. Sun to light shade, most soils, occasional to infrequent summer water. Dormant in summer if grown dry. Can grow in shallow water. Spreads by rhizomes and by seed in moist or wet locations. USDA: 7–9 Sunset: 4–9, 14–24 WUCOLS: L/M

Juniperus

JUNIPER

Evergreen coniferous shrubs and small trees with needlelike juvenile leaves, scalelike mature leaves, and berrylike cones. Some retain needlelike leaves into maturity. Sun, good to excellent drainage, infrequent to no summer water. WUCOLS: L/M

J. communis, juniper, variable, prostrate and widely spreading to tall and treelike, with green, needlelike mature leaves. Native to dry, open areas and forests or woodlands in temperate climates throughout the northern hemisphere. Most plants offered are cultivars. 'Point Saint George', a prostrate selection from coastal northern California, is 6 inches tall and 3–4 feet wide with silvery green leaves. Drapes over walls. USDA: 4–10 Sunset: 1–24

J. conferta, shore juniper, prostrate to 1–2 feet tall and 6–8 feet wide, with blue-green, prickly, needlelike mature leaves. Drapes over walls. Native to sandy soils in coastal areas of Japan. 'Blue Pacific' was selected for its dense, blue foliage and cold hardiness. May be offered as a variety of *J. rigida*. Good choice for coastal locations. USDA: 6–9 Sunset: 3–9, 14–24

J. scopulorum, Rocky Mountain juniper, upright and pyramidal to 25–35 feet tall and 10–15 feet wide, with dark green to bluish green, needlelike juvenile leaves and scalelike mature leaves. Native to dry, rocky slopes from sea level to 9,000 feet from southern British Columbia to Oregon and southeast through the Rocky Mountains to Arizona and Texas. 'Tolleson's Weeping', 15–20 feet tall and 8–12 feet wide, has blue-green leaves that drape from pendulous branches. USDA: 4–9 Sunset: 1–24

^ *Juniperus communis* 'Point Saint George'

Justicia

JUSTICIA

Evergreen and deciduous shrubs with showy, tubular flowers. Native to tropical, warm temperate, and semi-arid climates of the Americas, India, and Africa. USDA: 9–11

J. californica, chuparosa, semi-evergreen, 4–6 feet tall and wide, with small, light green leaves and bright red flowers in spring. Native to the deserts of southern California, Arizona, and Baja California. Sun to light shade, fast drainage, infrequent to no summer water. Drops leaves in summer if grown dry. May be offered as *Beloperone*. Sunset: 10–14, 18–24 WUCOLS: L/VL

^ *Justicia californica*

J. spicigera, Mexican honeysuckle, evergreen shrub, 3–4 feet tall and wide, with large, oval to broadly lance-shaped, light green leaves and orange flowers in spring through fall. Native to dry and moist forests and woodlands from sea level to almost 10,000 feet and from Mexico south to Colombia. Best in part shade with good drainage and moderate to occasional summer water. Sunset: 12–24 WUCOLS: L

^ *Justicia spicigera*

Keckiella

BUSH PENSTEMON

Evergreen shrubs with glossy, green leaves and tubular or trumpet-shaped flowers. Sun to part shade, most well-drained soils, good air circulation, occasional to infrequent or no summer water. Deciduous and dormant by late summer if grown dry. Cut back in late summer or fall to renew. May be offered as *Penstemon*. USDA: 7–10 Sunset: 7–9, 12–24

K. antirrhinoides, yellow bush penstemon, 3–5 feet tall and wide, with large, fragrant, yellow flowers in late winter or early spring. Native to dry, rocky slopes and mesas in chaparral and open woodlands of southern California, Arizona, and Baja California. WUCOLS: L

K. cordifolia, climbing or heartleaf penstemon, 3–6 feet tall and wide, with unscented, orange to red flowers in spring and summer. Native to rocky, well-drained slopes, often in shade, in coastal mountains of central and southern California and northern Baja California. WUCOLS: L/VL

^ *Keckiella cordifolia*

Koeleria macrantha

JUNE GRASS

Cool-season bunchgrass, variable, from less than 1 foot to 2 feet tall and 1–2 feet wide, with narrow, light green to bluish green leaves and silvery green flowers that age to tan. Self-sows freely. Native to dry, rocky soils, usually between 4,000 and 8,000 feet, in temperate regions of Europe, Asia, and North America. In North America it is found from southeastern Alaska to northern Mexico and east to the central East Coast. 'Barkoel' and 'Barleria' are cultivars of European plants offered as rough, low-maintenance turfgrasses. Best along the coast or at higher elevations. Sun to light shade, well-drained soil, moderate to occasional summer water. May be offered as a variety of *K. cristata*. USDA: 3–9 Sunset: N/A WUCOLS: L

Koeleria macrantha >

Lagerstroemia
CRAPE MYRTLE

Deciduous trees and shrubs with showy clusters of summer flowers and good fall color. Most plants offered are selections of *L. indica*, a red-flowered species native to southeastern China, or hybrids of *L. indica* and *L. fauriei*, a white-flowered species from Japan. Best where summers are long and hot, but some perform acceptably in cooler climates. Some cultivars do best with regular summer water. Those listed here accept moderate to occasional summer water in full sun with good drainage and good air circulation. USDA: 7–10 Sunset: 7–10, 12–14, 18–21 WUCOLS: L/M

Among mildew-resistant hybrids are *L.* 'Wichita' and *L.* 'Muskogee', both with lavender flowers, and *L.* 'Natchez', with white flowers, all 25–30 feet tall and 12–15 feet wide. *L.* 'Hopi', with pink flowers, and *L.* 'Acoma', with white flowers, both 8–10 feet tall and wide, also have good mildew resistance. *L.* 'Tuscarora' is less cold hardy than some other mildew-resistant cultivars.

Among plants most accepting of cold are two *L. indica* selections: PINK VELOUR ('Whit III'), 8–12 feet tall and wide, with hot pink flowers and burgundy leaves, and DYNAMITE ('Whit II'), 15–20 feet tall and 8–12 feet wide, with green leaves and red flowers.

Laurus nobilis
BAY LAUREL

Evergreen tree or large shrub, multitrunk, slow growing to 15–30 feet tall and 12–20 feet wide, with leathery, aromatic, dark green leaves and inconspicuous greenish yellow flowers in spring. Native to low-elevation and coastal habitats in the Mediterranean region. 'Crispa' has wavy-edged leaves. Cool sun to light shade, well-drained soils, moderate to occasional or infrequent summer water. Good formal or informal screen. USDA: 8–10 Sunset: 5–9, 12–24 WUCOLS: L

^ *Lagerstroemia indica*

^ *Laurus nobilis*

^ *Lavandula ×allardii* 'Meerlo'

^ *Lavandula angustifolia*

Lavandula

LAVENDER

Evergreen shrubs or shrubby perennials with aromatic, linear, lobed, or deeply dissected, green or gray-green leaves and showy spikes of small flowers, some with large, colorful bracts. Almost four dozen species and hundreds of named hybrids and selections. Sun to light shade, good to excellent drainage, good air circulation, occasional to infrequent summer water. WUCOLS: L/M

L. ×*allardii* 'Meerlo', 2–3 feet tall and wide, with gray-green leaves edged with creamy white and large, pale lavender flowers. Sport, discovered in New Zealand, of *L.* ×*allardii*, a hybrid believed to involve *L. dentata* and *L. latifolia*. USDA: 9–10 Sunset: 8–9, 12–24

L. 'Ana Luisa', 2–3 feet tall and wide, with woolly, silvery gray, almost white leaves and dark bluish purple flowers in early summer. A selection of *L.* ×*chaytoriae*, which is a hybrid involving *L. angustifolia* and *L. lanata*, a white-leaved species from southern Spain. Good choice for the Pacific Northwest. USDA: 6–9 Sunset: N/A

L. angustifolia, English lavender, 1–3 feet tall and 2–4 feet wide, with long, narrow, gray-green or silvery gray leaves and lavender, purple, pink, or white flowers in summer. Native to dry, rocky, exposed slopes in the mountains of eastern Spain to Italy, this is the most cold-hardy lavender. 'Hidcote' ('Hidcote Superior', 'Hidcote Blue'), with dark bluish purple flowers, and 'Munstead', with rosy purple flowers, are especially fragrant. Needs some winter chill to flower well. May be offered as *L. officinalis*. USDA: 5–10 Sunset: 2–24

L. dentata, French lavender, 2–4 feet tall and 4–5 feet wide, with narrow, gray-green leaves with scalloped margins and pale bluish purple flowers on tall stems in early spring through summer. The variety *candicans* has grayer leaves. Native to southern Spain and northern Africa. USDA: 8–10 Sunset: 8–9, 12–24

L. 'Goodwin Creek Grey', 2–3 feet tall and 3–4 feet wide, with densely woolly, silvery gray-green leaves and dark purple flowers on tall stems. Hybrid of garden origin believed to be a cross between *L. dentata* and *L. lanata* known as *L.* ×*ginginsii*. May be offered as a selection of *L. dentata*. USDA: 5–9 Sunset: 8–9, 12–24

L. ×intermedia, lavandin, 2–3 feet tall and wide, with gray-green or silvery gray leaves and bluish purple flowers in mid- to late summer. Hybrid of *L. angustifolia* and *L. latifolia*, a closely related plant from Western Europe. Plants are larger than English lavenders and tend to bloom later than both Spanish and English lavenders. 'Grosso' has broad spikes of especially fragrant, dark purple flowers on long stems. PHENOMENAL ('Niko'), a sport of 'Grosso', may be especially hardy. USDA: 6–9 Sunset: 4–24

L. stoechas, Spanish lavender, 2–3 feet tall and wide, with gray to blue-gray, woolly leaves and short, stocky spikes of dark purple flowers topped with large, pink or purple bracts from early to midspring into summer. Self-sows readily and can be weedy. Native to the Mediterranean region from France to Greece and Spain to Morocco. Earlier blooming and less hardy than both English lavender and *L.* ×*intermedia*. 'Otto Quast', with prominent, dark purple bracts, blooms almost year-round in mild climates. USDA: 7–9 Sunset: 4–24

Lavatera, see *Malva assurgentiflora*

^ *Lavandula dentata*

^ *Lavandula stoechas*

Layia

TIDYTIPS

Annuals with yellow or white daisy flowers in late winter to early spring. Sun to part shade, most soils, no summer water. Self-sow. Best grown from seed sown in fall. Need moisture in spring to bloom well. USDA: N/A WUCOLS: N/A

L. glandulosa, white tidytips, 1 foot tall and wide, with narrow, linear to lance-shaped, hairy, green leaves and white or sometimes pale yellow flowers with yellow centers. Native to sandy or gravelly soils in dry, open places, mostly inland, from British Columbia south to Baja California and east to Utah and Arizona. Sunset: N/A

Layia glandulosa >

^ *Layia platyglossa*

L. platyglossa, coastal tidytips, less than 1 foot tall and wide, with narrow, hairy, gray-green leaves and lightly fragrant, white-tipped, yellow flowers on tall stems. Native to low-elevation, dry habitats in the Central Valley and coastal mountains of central and southern California north to southwestern Oregon and south to northern Baja California. Sunset: 1–10, 14–24

Leonotis

LION'S TAIL

Evergreen shrubs and shrubby perennials with lance-shaped or oval to rounded, green leaves and whorled clusters of velvety, bright orange, tubular summer flowers encircling the stems at intervals. Cool sun, afternoon shade inland, well-drained soils, occasional summer water. USDA: 9–11 Sunset: 8–24 WUCOLS: L

^ *Leonotis leonurus*

L. leonurus, lion's tail, stiffly upright to 4–6 feet tall and 3–4 feet wide, with narrow, lance-shaped, dark green leaves with finely serrated margins. Native to rocky soils in grasslands of the Cape region of South Africa.

L. menthifolia*, see *L. ocymifolia

L. ocymifolia, mint-leaf lion's tail, 2–3 feet tall and wide, with small, rounded to oval, green leaves with scalloped margins. Native to rocky outcrops in eastern Africa from Sudan south to northeastern South Africa. May be offered as *L. menthifolia*.

^ *Leonotis ocymifolia*

Lepechinia
PITCHER SAGE

Evergreen shrubs with aromatic, softly hairy or felted leaves, stems, and bracts and tubular flowers at branch ends in spring or summer. Full sun along the coast, light shade inland, most well-drained soils, occasional to infrequent or no summer water. Sunset: 7–9, 14–24

L. calycina, white pitcher sage, 5–8 feet tall and 4–6 feet wide, with lance-shaped to oval, gray-green to yellow-green leaves and lavender-tinted white flowers in spring. Drops leaves in summer if grown dry. Native to rocky slopes and ridges along the coast and in coastal mountains of central to southern California and the northern Sierra Nevada foothills. 'Rocky Point' is 2–3 feet tall and 3–5 feet wide. USDA: 8–10 WUCOLS: L/M

^ *Lepechinia fragrans*

L. fragrans, island pitcher sage, 3–5 feet tall and wide, with oval to almost triangular, gray-green leaves and pinkish lavender flowers in spring and summer. Native to rocky, often north-facing slopes in the mountains of coastal southern California and the Channel Islands. USDA: 9–10 WUCOLS: L

L. hastata, Baja pitcher sage, 4–6 feet tall and wide, with almost triangular, felted, silvery green leaves and magenta flowers in summer and early fall. Native to Baja California and Baja California Sur. No summer water along the coast. USDA: 9–10 WUCOLS: L/M

Leptospermum
TEA TREE

Evergreen shrubs to small trees with small, needle-like or narrowly oblong to oval, aromatic, green to gray-green leaves and masses of small spring or early summer flowers, followed by persistent, woody seed capsules. Fast growing. Sun to part shade, most soils, occasional to infrequent or no summer water.

L. 'Dark Shadows', 12–15 feet tall and 15–20 feet wide, with burgundy-tinted, dark olive-green, lance-shaped leaves and white flowers. Believed to be a seedling of *L. macrocarpum* 'Copper Spray'. USDA: 9–10 Sunset: 14–24 WUCOLS: L

L. laevigatum, Australian tea tree, 15–30 feet tall and 15–20 feet wide, with grayish green, narrowly oblong leaves and white flowers. Native to sand dunes and bluffs in coastal southeastern Australia from New South Wales to Tasmania. 'Reevesii' ('Compacta') is 3–5 feet tall and 4–6 feet wide. Species is considered high risk for invasiveness in parts of coastal California. USDA: 9–11 Sunset: 14–24 WUCOLS: L

^ *Leptospermum lanigerum* 'Silver Form'

^ *Leptospermum scoparium* (Nanum Group) 'Tui'

L. lanigerum, woolly tea tree, variable, 8–15 feet tall and 6–12 feet wide, with softly downy, oblong, green leaves, silvery gray-green new growth, and white flowers. Native to mountains of southeastern Australia from New South Wales to South Australia and Tasmania, often along streams or in seasonally flooded places. Best in cool-summer climates. 'Silver Form', 6 feet tall and 5 feet wide, has tiny, almost pure silver leaves and white flowers. USDA: 8b–10 Sunset: N/A WUCOLS: N/A

L. namadgiensis, alpine tea tree, 4–6 feet tall and wide, with slightly woolly, oblong, gray-green to silvery green leaves and white or pale pink flowers. Bright orange bark peels to reveal green underbark. Native to higher elevations in the mountains southwest of Canberra in southeastern Australia. Good choice for the Pacific Northwest. USDA: 7b–9 Sunset: N/A WUCOLS: N/A

L. rupestre, alpine tea tree, variable, flat mat to 3–5 feet tall and 5–8 feet wide, with dark gray-green, oval leaves and white flowers. Lowest forms hug contours. Native to the mountains of southeastern Australia, especially Tasmania. 'Squiggly', upright, 5–6 feet tall and 3 feet wide, has dark red stems, silvery gray leaves, and masses of white flowers from pink buds. USDA: 7b–9 Sunset: N/A WUCOLS: L/M

L. scoparium, New Zealand tea tree, variable, from prostrate to 30 feet tall, with narrowly linear, prickly tipped, green or gray-green leaves and white, pink, or red, single or double flowers. Native to most of New Zealand and to southern Australia from Western Australia to New South Wales. 'Star Carpet' is less than a foot tall and 6 feet wide, with dark green, sharp-tipped, scalelike leaves and single white flowers. 'Washington Park', 8 feet tall and 4 feet wide, with white flowers, is a cold-hardy selection from the mountains of New Zealand, named for and growing in the Washington Park Arboretum in Seattle. (Nanum Group) 'Tui', 4–6 feet tall and wide, has pale pink flowers and cascades over rocks or walls. USDA: 8–11 Sunset: 14–24 WUCOLS: M

Leptosyne gigantea
GIANT COREOPSIS

Perennial, 3–5 feet tall and wide, with finely dissected, bright green leaves atop a leafless, woody stem and bright yellow flowers from late winter to late spring. Deciduous in summer if grown dry. Native to sand dunes and coastal bluffs in southern California, including the Channel Islands. Sun, well-drained soils, no summer water. Good seaside plant. May be offered as *Coreopsis*. USDA: 9–10 Sunset: N/A WUCOLS: L/VL

^ *Leptosyne gigantea*

^ *Lessingia filaginifolia*

Lessingia filaginifolia

CALIFORNIA ASTER

Perennial, less than 1 foot tall and 3–6 feet wide, with small, silvery gray-white leaves and lavender, pink, or white flowers with yellow centers in late spring and summer. Native to sandy or rocky soils on coastal bluffs and mountain slopes from southwestern Oregon to Baja California. 'Silver Carpet', from Monterey County, has pinkish lavender flowers. Sun to part shade, most well-drained soils, occasional to infrequent summer water. Good along the coast or inland. May be offered as *Corethrogyne*. USDA: 8b–10 Sunset: N/A WUCOLS: L/M

Leucadendron

CONEBUSH

Evergreen shrubs and small trees with lance-shaped, elliptical, or sometimes needlelike leaves and colorful bracts surrounding conical flowerheads that form woody, conelike seedheads after flowering. Most bloom in winter or spring. Native to South Africa. Many leucadendrons are difficult in cultivation, needing perfect drainage, acidic soils, specifically targeted fertilization, and careful, consistent, sometimes regular summer water. Listed here are a few more adaptable cultivars that thrive in cool sun with excellent drainage and occasional or infrequent summer water. USDA: 9b–11 Sunset: 16–17, 20–24 WUCOLS: L/M

^ *Leucadendron* 'Safari Sunset'

^ *Leucophyllum frutescens*

^ *Leucophyllum langmaniae*

L. 'Ebony', 3–4 feet tall and 3–6 feet wide, with lance-shaped, dark purple, almost black leaves and red bracts. Sport of *L.* 'Safari Sunset' discovered in New Zealand in a cultivated cut-flower field.

L. 'Jubilee Crown', 4–6 feet tall and 3–5 feet wide, with fine-textured, needlelike, gray-green leaves and pink-tinged green bracts. Hybrid involving *L. laxum* or *L. lanigerum*, possibly both.

L. 'Safari Sunset', 8–10 feet tall and 6–8 feet wide, with lance-shaped, green leaves and dark red bracts that take on creamy yellow tones in spring. Hybrid between *L. salignum* and *L. laureolum*.

***L. salignum* 'Winter Red'**, 3–4 feet tall and 4–6 feet wide, with lance-shaped, green leaves flushed purplish red in cool weather, and red and creamy white bracts.

Leucophyllum

TEXAS RANGER

Evergreen shrubs with felted, silvery gray or green leaves and lightly fragrant, funnel-shaped flowers in late summer and early fall. Full sun, excellent drainage, occasional to infrequent or no summer water. Best in hot-summer, mild-winter climates and alkaline soils. USDA: 8–10 Sunset: 7–24 WUCOLS: L

L. frutescens, Texas ranger, 5–8 feet tall and wide, with silvery gray leaves and pinkish purple flowers. Native to rocky soils in the Chihuahuan Desert from northeastern Mexico to southern Texas and New Mexico. GREEN CLOUD has green leaves and purple flowers. 'White Cloud' has pale gray leaves and white flowers.

L. langmaniae, 4–5 feet tall and wide, with sage green leaves and lavender flowers. Native to the Chihuahuan Desert in northern Mexico. 'Lynn's Legacy' ('Lynn's Everblooming') is a long-flowering selection.

^ *Limnanthes douglasii*

^ *Leucophyta brownii*

Leucophyta brownii
CUSHION BUSH

Evergreen shrub, 1–3 feet tall and 2–3 feet wide, with tiny, silvery white, scalelike leaves and small, button-shaped heads of tiny, pale yellow to creamy white flowers in spring and summer. Native to coastal dunes and cliffs along the south coast of Australia from southwestern Western Australia to the north coast of Tasmania. Sun, fast drainage, infrequent summer water. Good seaside plant. May be offered as *Calocephalus*. USDA: 9–10 Sunset: N/A WUCOLS: L

Leymus, see *Elymus*

Limnanthes douglasii
MEADOWFOAM

Annual, less than 1 foot tall and 1 foot wide, with finely divided, medium green leaves and lightly fragrant spring flowers, petals usually yellow with white tips but sometimes all yellow or all white. Self-sows freely. Native to the edges of vernal pools and seasonally moist meadows at low elevations in coastal mountains and valleys from central California to northern Oregon. Sun, moisture-retentive soils, no summer water. USDA: N/A Sunset: 1–9, 14–24 WUCOLS: N/A

^ *Lomandra hystrix* KATIE BELLES

^ *Lomandra hystrix* TROPIC BELLE

^ *Lomandra longifolia*

Lomandra

MAT RUSH

Perennials from rhizomes, clump-forming, with grasslike or narrowly straplike leaves and flowering stems, usually shorter than the leaves, bearing clusters of tiny, whitish yellow spring or summer flowers and noticeably bristly bracts. Native to varied habitats in Australia, from semi-arid to rainforest. Sun to part shade, most soils, moderate to occasional summer water. May spread by rhizomes in moist soil. Cut back in spring every few years to renew. Sunset: N/A

***L. confertifolia* subsp. *rubiginosa* 'Seascape'**, 1–2 feet tall and 2–3 feet wide, with narrow, dark blue-gray leaves that arch over to the ground. Subspecies is native to open forest in New South Wales. USDA: 8b–11 WUCOLS: M

***L. fluviatilis* SHARA ('AU807')**, a little more than 1 foot tall and 2 feet wide, with narrow, grayish or bluish green, arching leaves. Species is native to moist soils in New South Wales but thrives with occasional summer water. USDA: 9–11 WUCOLS: N/A

L. hystrix, 3–5 feet tall and wide, with medium-width, green leaves. Accepts periodic dry soils but prefers occasional to moderate summer water. Native to moist sites in Queensland and New South Wales. KATIE BELLES ('LHBYF'), 3–5 feet tall and wide, has bright green leaves and prefers part shade. TROPIC BELLE ('LHCOM'), 2–3 feet tall and 3–4 feet wide, has yellow-green leaves. USDA: 8–11 WUCOLS: M

L. LIME TUFF ('Lomlon' or 'Bushland Green'), dwarf mat rush, upright to 2–3 feet tall and wide, with narrow, deep green leaves. Hybrid of *L. longifolia* and *L. confertifolia* selected in Australia. Good in sun or shade. USDA: 8–11 WUCOLS: N/A

L. longifolia, spiny-headed mat rush, 3–5 feet tall and wide, variable plant with straplike leaves. Succeeds in wet or dry soils and in sun or shade. Native to eastern Australia. BREEZE ('LM300' or

'Tanika'), 2–3 feet tall and wide, with narrower, dark green leaves, may be more cold hardy. EVERGREEN BABY ('LM600'), 18 inches tall and wide, has even narrower leaves. PLATINUM BEAUTY ('Roma13'), 2–3 feet tall and wide, has green leaves with a white center stripe and white edges. USDA: 8–10 WUCOLS: L/M

Lonicera

HONEYSUCKLE

Deciduous and evergreen shrubs, vines, and vining shrubs with oval, green to bluish green leaves and clusters of fragrant, trumpet-shaped spring and summer flowers, followed by red or orange to blue-purple berries. Several honeysuckles are invasive, including *L. japonica*, from eastern Asia, and many need regular summer water. Those listed here are not known to be invasive and thrive with moderate to infrequent summer water in most soils and in cool sun or part shade. Not low water in hot-summer climates.

L. ciliosa, western trumpet honeysuckle, deciduous vining shrub, to 10–20 feet, with red-orange flowers and orange berries. Native to north-facing slopes and shaded streambanks in dry to moist forests from southwestern British Columbia to northwestern California, mostly west of the Cascades. Best in northern California and the Pacific Northwest. Moderate summer water. USDA: 5–9 Sunset: N/A WUCOLS: M

L. hispidula, hairy honeysuckle, deciduous to semi-evergreen vining shrub, to 6–10 feet, with lavender to bright pink flowers and red berries. Native to south- or west-facing slopes and forest openings, often along streambanks, from southwestern British Columbia to southern California. Best along the coast with some afternoon shade and infrequent summer water. USDA: 6–10 Sunset: N/A WUCOLS: L/VL

L. nitida, boxleaf honeysuckle, evergreen shrub, fast growing to 8–10 feet tall and wide, with creamy white, late spring flowers and blue-purple berries.

^ *Lonicera hispidula*

^ *Lonicera nitida* 'Twiggy'

Native to sunny or partly shaded streamsides at 4,000 to 9,800 feet in the mountains of southwestern China. 'Baggesen's Gold', 4–6 feet tall and wide, has bright yellow new leaves. 'Twiggy', a sport of 'Baggesen's Gold', is 2–3 feet tall and wide. USDA: 7–9 Sunset: 4–9, 14–24 WUCOLS: M

L. subspicata, southern or chaparral honeysuckle, evergreen vining shrub, to 3–8 feet tall, with creamy white to pale yellow flowers and red or yellow berries. Native to dry, west-facing slopes, usually partly shaded by other vegetation, in coastal mountains of central and southern California and northern Baja California. USDA: 7–10 Sunset: N/A WUCOLS: L

Lupinus

LUPINE

Annuals and perennials with bluish green leaves divided into many leaflets and dense spikes of

flowers in spring. Self-sow. Many hybrid lupines, commonly available in garden centers, need regular water. Those listed here are content with infrequent to no summer water in sun to part shade and most well-drained soils. Need ample moisture in spring to bloom well. *L. arboreus*, yellow bush lupine, is invasive in some parts of California.

L. albifrons, silver bush lupine, perennial, 3–5 feet tall and wide, with silvery white leaves and upright spikes of blue and purple flowers. Native to rocky soils in meadows and forest openings from southern Oregon to northern Baja California. USDA: 8–10 Sunset: N/A WUCOLS: L/VL

L. nanus, sky lupine, annual, less than 1 foot to 2 feet tall and 1 foot wide, with blue and white flowers. Native to woodland openings and grasslands in coastal central California inland to the Sierra Nevada foothills and east to Nevada and eastern Oregon. USDA: N/A Sunset: 3–24 WUCOLS: N/A

L. succulentus, arroyo lupine, annual, 2–3 feet tall and wide, with blue and purple flowers. Self-sows freely. Native to seasonally moist places in much of California, western Arizona, and northern Baja California. USDA: N/A Sunset: 7–24 WUCOLS: N/A

^ *Lyonothamnus floribundus* subsp. *aspleniifolius*

Lyonothamnus floribundus subsp. *aspleniifolius*

SANTA CRUZ ISLAND IRONWOOD

Evergreen tree, 25–40 feet tall and 15–25 feet wide, with large, deeply divided and notched, dark green leaves, paler beneath, and clusters of creamy white flowers in late spring to summer, followed by persistent seedpods. Fibrous, gray bark peels to reveal shiny, red underbark. Native to rocky soils on north-facing slopes on the Channel Islands in southern California. Cool sun to part shade or afternoon shade inland, excellent drainage, occasional to infrequent summer water. Best near the coast and in mild-winter climates. USDA: 8–10 Sunset: 14–17, 19–24 WUCOLS: L

< *Lupinus albifrons*

^ *Madia elegans*

^ *Mahonia eurybracteata*

^ *Maireana sedifolia*

Madia elegans

TARWEED

Annual, 2–3 feet tall and 1–2 feet wide, with aromatic, green leaves covered in sticky hairs, and bright yellow daisy flowers on leafy, 4- to 5-foot stems from midsummer to fall. Self-sows. Native to dry slopes, grasslands, and open woodlands below 3,300 feet from southwestern Washington to northern Baja California. Flowers close during midday heat, opening late in the afternoon. Sun to part shade, most soils, occasional to infrequent summer water. USDA: N/A Sunset: N/A WUCOLS: N/A

Mahonia aquifolium, M. nevinii, M. repens, see *Berberis*

Mahonia eurybracteata

THREADLEAF MAHONIA

Evergreen shrub, 3–5 feet tall and wide, fine-textured, with foot-long leaves divided into narrow, linear, dark green leaflets, bright yellow flowers in fall or early winter, and silvery dark blue berries. Unlike most other mahonias, leaflets have no spines. Native to southwestern China. Usually available as 'Soft Caress'. Part shade to shade, well-drained, humusy soils, occasional to infrequent or no summer water. May be offered as *M. confusa*. USDA: 7b–11 Sunset: 6–9, 14–24 WUCOLS: N/A

Maireana sedifolia

PEARL BLUEBUSH

Evergreen shrub, upright to 3–4 feet tall and wide, with grayish, nearly white, semi-succulent leaves and inconspicuous pinkish white flowers. Native to inland, arid and semi-arid parts of South Australia. Sun to light shade, most soils, occasional to infrequent or no summer water. USDA: 9–11 Sunset: N/A WUCOLS: L

Malacothamnus
BUSH MALLOW

Evergreen shrubs, upright and spreading widely by underground stems to form dense thickets. Oval to rounded, deeply or shallowly lobed, green to gray-green, felted leaves and elongated clusters of upfacing, loosely cup-shaped, pink or pale lavender flowers in spring and summer. Stems, leaves, and buds densely covered in white or brownish white hairs lend a soft, gray or silvery appearance. Sun to part shade, most well-drained soils, infrequent to no summer water. Best with some summer heat. Good bank covers. May be short lived. Sunset: N/A WUCOLS: L/VL

M. fasciculatus, chaparral mallow, variable, 4–12 feet tall. Native to the San Francisco Bay Area, southern California, and Baja California, mostly near the coast but also inland. May be offered as *M. arcuatus*. USDA: 8–11

M. fremontii, Fremont's bush mallow, stiffly upright to 6–8 feet tall. Native to interior Coast Ranges in northern California and Sierra Nevada foothills. Good choice for California's Central Valley. USDA: 8–10

M. jonesii, slender bush mallow, upright, 4–8 feet tall. Native to lower elevations of coastal mountains and valleys of Monterey and San Luis Obispo counties in central California. USDA: 9b–11

^ *Malacothamnus fasciculatus*

^ *Malosma laurina*

Malosma laurina
LAUREL SUMAC

Evergreen shrub to small tree, fast growing to 8–20 feet tall and 15–20 feet wide, with large, lance-shaped, aromatic, green leaves with red margins and large clusters of tiny, fragrant, white flowers in late spring and early summer. Native to sandy or rocky slopes and bluffs below 3,000 feet in coastal mountains and valleys from Point Conception and the Channel Islands to Baja California. Sun to part shade, good drainage, infrequent to no summer water. May be offered as *Rhus*. USDA: 9–10 Sunset: N/A WUCOLS: L/VL

Malva assurgentiflora
ISLAND MALLOW

Evergreen shrub, fast growing to 4–12 feet tall and 5–10 feet wide, with large, dark green, maplelike leaves and showy, pinkish purple flowers with darker purple veins from midspring to late summer. Native to coastal bluffs on the Channel Islands in southern California and naturalized on the coastal mainland. 'Purisima', a hybrid with *M. venosa* from the San Benito Islands of Baja California, is similar but with maroon flowers. Sun to part shade, most well-drained soils, occasional to infrequent or no summer water. May be offered as *Lavatera*. USDA: 8–10 Sunset: 14–24 WUCOLS: L/M

^ *Malva assurgentiflora* 'Purisima'

Marrubium bourgaei
HOREHOUND

Perennial, 1–2 feet tall and 2–4 feet wide, with rough-textured, lime-green leaves and small, greenish to creamy white flowers in spring. Native to rocky soils in mountains of Turkey and parts of western Asia. Usually available as 'All Hallows Green'. Sun to light shade, most well-drained soils, occasional summer water. May be offered as *Ballota*. USDA: 8–9 Sunset: N/A WUCOLS: L/M

^ *Marrubium bourgaei* 'All Hallows Green'

Melaleuca
MELALEUCA

Evergreen trees and shrubs with narrow, needlelike or linear leaves and clusters of brushlike flowers in spring or summer, followed by persistent, woody seed capsules. Full sun, most soils, occasional to infrequent or no summer water. Good seaside plants. Can be weedy or invasive.

M. citrinus*, see *Callistemon

M. elliptica, granite honey myrtle, 8–10 feet tall and 6–8 feet wide, with small, linear to oval, gray-green leaves and red flowers in late spring to early summer. Native to rocky outcrops in southern Western Australia. USDA: 9–10 Sunset: N/A WUCOLS: L/VL/M

M. incana, gray honey myrtle, fast growing to 6–10 feet tall and 8–12 feet wide, with needlelike, lightly hairy, gray-green leaves on pendulous branches and creamy yellow flowers in spring. Native to southwestern Western Australia. A low groundcover form is often available. USDA: 9–11 Sunset: 8–9, 12–24 WUCOLS: L/M

Melaleuca incana >

M. linariifolia, flaxleaf paperbark, 20–30 feet tall and 15–25 feet wide, with stiff, gray-green, narrowly linear leaves, papery white, peeling bark, and white flowers in summer. Native to dry forests, usually near streams, in coastal Queensland and New South Wales. Accepts both dry and moist conditions. USDA: 10–11 Sunset: 9, 13–24 WUCOLS: L

M. nesophila, pink melaleuca, 15–20 feet tall and 10–15 feet wide, with small, linear to oval, bright green leaves, spongy, peeling, grayish white bark, and lavender flowers for much of the year. Native to the southern coast of Western Australia. USDA: 9–11 Sunset: 13, 16–24 WUCOLS: L

M. quadrifida, see Calothamnus

Melampodium leucanthum

BLACKFOOT DAISY

Perennial, 6–12 inches tall and 1–2 feet wide, with narrow, dark green, hairy leaves and bright white daisy flowers with yellow centers from spring to fall. Self-sows freely. Native to dry, rocky soils from southern Colorado to central Texas and south to northern Mexico. Sun to part shade, fast drainage, occasional summer water. Needs summer heat. Cut back in late winter to renew. USDA: 6–11 Sunset: 2–3, 10–13 WUCOLS: L

^ *Melampodium leucanthum*

Melica

MELIC GRASS

Cool-season bunchgrasses with narrow leaves and small, airy spring flowers with translucent, papery bracts. Most readily self-sow. Cool sun to part shade, most well-drained soils, infrequent to no summer water. Dormant by midsummer if grown dry. Sunset: N/A

^ *Melica californica*

M. californica, California melic, 1 foot tall and wide, with bright green leaves and narrow flower spikes on 2- to 3-foot stems. Native to dry, exposed, rocky slopes below 6,000 feet in the Coast Ranges and Sierra Nevada foothills of central to northern California and southwestern Oregon. Best in part shade with occasional summer water. USDA: 7–10 WUCOLS: L/VL

M. imperfecta, coast melic, 1 foot tall and wide, with narrow, green to yellowish green leaves and wiry, 2- to 3-foot flowering stems. Native to dry, rocky hillsides from sea level to 4,000 feet in the southern Sierra Nevada, central to southern California coastal mountains, the western Mojave desert, and northern Baja California. Infrequent to no summer water. USDA: 7–10 WUCOLS: VL

M. torreyana, Torrey's melic, 1–2 feet tall and wide, with fine-textured, grayish green leaves and purplish white seedheads. Native to chaparral and forests below 3,000 feet in central California coastal mountains and Sierra Nevada foothills. Best in cool sun to part shade with infrequent to no summer water. USDA: 8–10 WUCOLS: VL

^ *Melica imperfecta*

Mentzelia lindleyi

BLAZING STAR

Annual, 1–2 feet tall and wide, with narrow, deeply lobed, green leaves and glossy, fragrant, bright yellow, star-shaped flowers on 2- to 3-foot stems in late spring or early summer. Flowers open in the late afternoon and close with the morning sun. Native to dry, rocky slopes and open brushy areas below 4,000 feet in the Coast Ranges of central California and the northern Sierra Nevada foothills east to Arizona. Full sun, excellent drainage, infrequent to no summer water. Needs summer heat. USDA: N/A Sunset: N/A WUCOLS: N/A

^ *Mentzelia lindleyi*

Mimulus, see *Diplacus*

Monardella

COYOTE MINT

Annuals and perennials with aromatic leaves and clusters of showy, tubular flowers in late spring to early summer. Sun to part shade, excellent drainage, occasional to infrequent or no summer water. Those listed here are perennials. Sunset: 7–9, 14–24

M. macrantha, scarlet monardella, 1 foot tall and 1–2 feet wide, with glossy, dark green leaves and loose clusters of large, bright red to red-orange flowers. Not easy but showy and unusual. Native to coastal mountains of central and southern California to Baja California, usually between 2,000 and 6,500 feet. *M. macrantha* subsp. *macrantha* 'Marian Sampson', from the mountains of Riverside County, seems particularly successful in gardens. Prefers cool sun or part shade with occasional summer water. USDA: 8–11 WUCOLS: L/M

M. villosa, coyote mint, 1–2 feet tall and 2–3 feet wide, with gray-green to blue-green leaves and spherical clusters of bright lavender or pinkish purple flowers. Native to rocky slopes and woodland openings below 4,000 feet in the Sierra Nevada foothills and Coast Ranges from central California to southern Oregon. *M. villosa* subsp. *franciscana* 'Russian River' is a selection from Sonoma County in northern California. Cut back in late winter to renew. USDA: 6–10 WUCOLS: L/VL

^ *Monardella macrantha* subsp. *macrantha* 'Marian Sampson' (red flowers) with *Salvia dorrii*

^ *Morella californica*

Morella californica

PACIFIC WAX MYRTLE

Evergreen shrub, usually multitrunk, upright to 10–20 feet tall and 8–10 feet wide, with aromatic, lance-shaped, glossy green leaves and inconspicuous flowers and fruits. Native to coastal and inland foothills and valleys from sea level to 2,000 feet from southwestern British Columbia to southern California. Sun along the coast, afternoon shade inland, most soils, moderate to occasional summer water. Good seaside plant. May be offered as *Myrica*. USDA: 7–10 Sunset: 4–9, 14–24 WUCOLS: L/M

Muhlenbergia

MUHLY, HAIR GRASS

Warm-season bunchgrasses with fine-textured leaves and tall plumes of flowers followed by airy seedheads in late summer and fall. Most self-sow. Sun or light shade, most well-drained soils, good air circulation, moderate to occasional or infrequent summer water.

^ *Muhlenbergia dubia*

M. capillaris, pink muhly, 2–3 feet tall and 4–5 feet wide, with blue-green leaves and a haze of purplish pink flowers. Native to grasslands and open woodlands from Massachusetts to Kansas south to Florida and west to Texas. 'White Cloud', possibly a selection of *M. capillaris* var. *filipes* (now *M. sericea*), is taller and more erect with white flowers. USDA: 7–11 Sunset: 4–24 WUCOLS: L/M

M. dubia, pine muhly, 1–2 feet tall and 2–3 feet wide, stiffly upright, with light green leaves and narrow, creamy white flowers on 3- to 4-foot stems. Native to eastern Arizona, New Mexico, southern Texas, and northern Mexico between 3,300 and 5,000 feet. USDA: 7–10 Sunset: 3b, 7–24 WUCOLS: L

M. lindheimeri, Texas muhly, upright to 3–5 feet tall and 4 feet wide, with medium green to blue-gray leaves and 5- to 6-foot stems with showy, purple-tinged flowers that age to silvery gray. Native to central Texas and northern Mexico. AUTUMN GLOW ('Leni') has creamy white flowers that age to pale yellow. USDA: 7–11 Sunset: 6–24 WUCOLS: L/M

M. 'Pink Flamingo', narrowly upright to 3–4 feet tall and 2–3 feet wide, with narrow, gray-green to blue-green leaves and pale pinkish gray flowers on 5-foot stems. Hybrid of garden origin from southeastern Texas, believed to be between *M. lindheimeri* and *M. capillaris*. USDA: 6–10 Sunset: N/A WUCOLS: N/A

M. reverchonii, rose or seep muhly, 1–2 feet tall and 2 feet wide, with fine-textured, medium green leaves and reddish pink flowers on 2- to 3-foot stems. Native to rocky slopes from central Oklahoma south to central Texas. UNDAUNTED ('PUND01S'), from north-central Texas, was selected for cold hardiness and performance in heat and heavy soils. USDA: 5–9 Sunset: N/A WUCOLS: N/A

M. rigens, deer grass, 2–3 feet tall and wide, with narrow, arching, bright green to gray-green leaves and yellow to purplish flowers on 5-foot stems. Native to many habitats below 7,000 feet from northern California south to Mexico and east to Texas. Good seaside grass. USDA: 7–9 Sunset: 4–24 WUCOLS: L/M

^ *Muhlenbergia reverchonii*

^ *Muhlenbergia rigens* (foreground)

Myrica, see *Morella*

Myrsine africana

AFRICAN BOXWOOD

Evergreen shrub, 4–8 feet tall and 3–6 feet wide, with small, aromatic, leathery, dark green leaves, new growth reddish, and inconspicuous yellow flowers in spring. Native to both summer- and winter-rainfall regions of east and southern Africa and parts of Asia. Sun to light shade, most soils, moderate to occasional or infrequent summer water. USDA: 9–11 Sunset: 8–9, 14–24 WUCOLS: L/M

^ *Myrsine africana*

^ *Myrtus communis* 'Variegata'

Myrtus communis
MYRTLE

Evergreen shrub, slow growing to 6–10 feet tall and wide, with small, aromatic, glossy, dark green leaves, new growth red, and small, fragrant, white flowers with long yellow stamens in summer. Native to Mediterranean regions of southern Europe, northern Africa, and western Asia. 'Andy's Hardy', 'Andy Van', and *M. communis* subsp. *tarentina* are smaller and said to be more cold hardy. 'Variegata' has green leaves with creamy white margins. Sun to light shade, well-drained soils, occasional to infrequent or no summer water. USDA: 8–10 Sunset: 8–24 WUCOLS: L/M

Narcissus
DAFFODIL

Perennials from bulbs, with strap-shaped to almost rushlike leaves and late winter or early spring flowers, singly or in small clusters. Flowers are yellow, white, or some combination of the two with petals surrounding a variously shaped central structure, or cup. Native to the western Mediterranean region, especially southwestern Spain and Portugal. 'Cragford', with pure white petals and a bright orange cup, is a good choice for warmer climates. Sun to light shade, most well-drained soils, no summer water. Need moisture in late winter to bloom well. USDA: 4–8 Sunset: 1–24, A2–3 WUCOLS: L/VL

^ *Narcissus* 'Cragford'

Nassella, see *Stipa*

Nepeta ×*faassenii*

CATMINT

Perennial, 1–2 feet tall and 2–3 feet wide, with aromatic, silvery gray-green leaves and upright spikes of lavender-blue flowers in summer. Hybrid of *N. racemosa*, native to the Caucasus, Turkey, and northern Iran, and *N. nepetella*, native from southern Spain to Italy, Morocco, and Algeria. 'Select Blue', 18 inches tall and wide, has darker blue-violet flowers. 'Six Hills Giant' and 'Walker's Low', both 2–3 feet tall and 4 feet wide, are especially vigorous and long blooming. Sun to part shade, good drainage, moderate to occasional summer water. 'Walker's Low' may be offered as *N.* 'Walker's Low' or as *N. racemosa* 'Walker's Low'. USDA: 4–8 Sunset: 1–24 WUCOLS: L/M

^ *Nepeta* ×*faassenii* 'Six Hills Giant' with iris and cotinus foliage

Nolina

BEAR GRASS, NOLINA

Succulent rosettes of grasslike or strap-shaped leaves with finely serrated margins but no spines and dense clusters of tiny, fragrant flowers on tall, branched stalks in spring. Native to desert regions with occasional summer rain. Sun to part shade, excellent drainage, occasional to infrequent or no summer water. Need summer heat. WUCOLS: L/VL

^ *Nepeta* ×*faassenii* 'Walker's Low' (foreground)

N. bigelovii, desert nolina, 3–6 feet tall and wide, with leathery, bluish gray-green leaves and creamy white flowers on a 6- to 12-foot stalk. Gradually forms a sturdy 3- to 6-foot, partially underground stem. Native to dry, rocky slopes and canyons from 500 to 5,000 feet in the deserts of southeastern California, southern Nevada, western Arizona, and northern Mexico. USDA: 7–10 Sunset: 7–16, 18–24

N. microcarpa, bear grass, 3–4 feet tall and 4–6 feet wide, with glossy, yellowish green leaves with fraying, curling tips and barely visible marginal teeth. Greenish white flowers on a 3- to 6-foot stalk. Short underground stem may spread to form new rosettes. Native to rocky slopes in desert grasslands and open woodlands from 3,000 to 6,500 feet in northern Mexico, Arizona, and New Mexico. USDA: 7–9 Sunset: 3, 10–13

N. nelsonii, blue nolina, 3–4 feet tall and wide, with stiff, silvery blue-green leaves with finely toothed margins and creamy white flowers on a 4- to 5-foot stalk. Gradually forms one to several 6- to 8-foot stems topped by individual rosettes. Native to deserts and mountains of northeastern Mexico. USDA: 8–10 Sunset: 5, 8–9, 11–24

^ *Nolina nelsonii*

N. parryi, giant nolina, is similar to *N. bigelovii* and may be offered as a subspecies. Native to dry slopes and ridges in the mountains and deserts of southern California, Arizona, and Baja California. USDA: 7–11 Sunset: 2–3, 7–24

^ *Nolina parryi*

Oemleria cerasiformis

OSOBERRY

Deciduous shrub or small tree, upright and thicket-forming to 8–20 feet tall and 6–15 feet wide, with medium green, lance-shaped leaves, lime green when new, and pendant clusters of small, fragrant, greenish white, trumpet-shaped flowers in late winter or early spring. Native to forest edges and openings west of the Cascades from sea level to 5,600 feet and from southern British Columbia to central California. Part shade to shade, most soils, moderate to occasional summer water. May be offered as *Osmaronia*. Not low water in hot-summer climates. USDA: 6–9 Sunset: 4–9, 14–24 WUCOLS: L/M

^ *Olea europaea*

Olea europaea

OLIVE

Evergreen tree, slow growing to 25–35 feet tall and wide, with gray-green, lance-shaped leaves, silvery beneath, small white flowers, and oval, green fruits. Native to southern Europe, western Asia, and northern Africa. Fruit drop can be messy and trees can be invasive. Fruitless cultivars have been developed, but all may produce a few fruits occasionally. MAJESTIC BEAUTY ('Monher') may produce the least fruit. Other essentially fruitless trees are 'Swan Hill' ('Arizona Fruitless') and 'Wilsonii'. LITTLE OLLIE ('Montra'), 6–8 feet tall, is fruitless and 'Skylark Dwarf', 10–15 feet tall, only occasionally sets a crop of small fruit. 'Frantoio', 'Arbequina', and 'Leccino', cold-hardy fruiting varieties from high-elevation Spain, are good performers in the Pacific Northwest. Sun, good drainage, infrequent to no summer water. USDA: 8–10 Sunset: 8–9, 11–24 WUCOLS: L/VL

^ *Olea europaea* LITTLE OLLIE

^ *Olea europaea* 'Wilsonii'

< *Oemleria cerasiformis*

^ *Olearia ×haastii*

^ *Origanum* 'Kent Beauty'

^ *Origanum laevigatum*

Olearia ×haastii

DAISY BUSH

Evergreen shrub, 4–6 feet tall and 6–8 feet wide, with small, glossy, dark green leaves, creamy white beneath, and clusters of fragrant white flowers in summer. Natural hybrid from the mountains of South Island, New Zealand, believed to be a cross of *O. moschata*, with small, gray leaves, and *O. avicenniifolia*, with large, bright green leaves. Sun to part shade, well-drained soils, infrequent to no summer water. Good seaside hedge or screen. USDA: 8–11 Sunset: N/A WUCOLS: N/A

Origanum

OREGANO

Perennials with small, green to gray-green, often aromatic leaves and clusters of tiny flowers with small bracts or inconspicuous flowers within elongating spikes of large, showy, overlapping bracts. Some spread by rhizomes or by rooting stems. Sun to part shade, fast-draining soils, occasional to infrequent summer water. WUCOLS: L/M

O. dictamnus, dittany of Crete, 6–12 inches tall and 1–2 feet wide, with arching spikes of rose-pink flowers and large, pale green, pink-tinged bracts. Native to rocky outcrops in the mountains and canyons of Crete. USDA: 7–10 Sunset: 8–9, 12–24

O. 'Kent Beauty', 6–10 inches tall and 18 inches wide, with large, pink bracts with a deep purple cast if grown in sun. Garden origin, likely a hybrid between *O. rotundifolium*, from Turkey, and *O. scabrum*, from the mountains of southern Greece, but commonly offered as a selection of the former. USDA: 6–9 Sunset: 2b–24

O. laevigatum, 1–2 feet tall and 2–3 feet wide, with loose clusters of tiny pink or purple flowers and small, darker purple bracts. Native to rocky soils in Turkey and Cyprus. 'Hopleys' ('Hopley's Purple'),

^ *Osmanthus delavayi*

developed in England, has dark gray-green leaves that turn purplish as the weather cools. USDA: 7–10 Sunset: 2–24

O. libanoticum, hop-flowered or cascading oregano, 1–2 feet tall and 2 feet wide, with magenta flowers and large, pink-tinted, pale chartreuse bracts on arching stems. Native to the mountains of Lebanon and Syria. Cascades over rocks and walls. USDA: 5–9 Sunset: 2b–24

O. rotundifolium, roundleaf oregano, 6–8 inches tall and 1–2 feet wide, with pink-tinged, white flowers and broad spikes of large, silvery chartreuse bracts. Native to the mountains of northeastern Turkey. USDA: 6–10 Sunset: 2b–24

Osmanthus

FALSE HOLLY, HOLLY OLIVE

Evergreen shrubs to small trees with leathery, dark green leaves, sometimes toothed or serrated, and clusters of tiny, fragrant, usually white flowers. Slow growing. Most are native to eastern Asia. Cool sun to part shade, most soils, moderate to occasional or infrequent summer water. WUCOLS: M

O. delavayi, tea olive, 6–8 feet tall and 8–10 feet wide, with tiny, finely toothed leaves and masses of white flowers in spring. Native to the mountains of western and southwestern China. USDA: 7–9 Sunset: 4–9, 14–21

^ *Osmanthus heterophyllus* 'Goshiki'

^ *Ozothamnus rosmarinifolius* 'Silver Jubilee'

O. ×fortunei, tea olive, 12–20 feet tall and 12–15 feet wide, with spiny-edged juvenile leaves, smooth-edged at maturity, and white flowers in late summer and fall. Hybrid of garden origin between *O. heterophyllus* and *O. fragrans*. 'San Jose', developed in California, has larger, creamy white to pale orange flowers. USDA: 7–9 Sunset: 4–10, 14–24

O. fragrans, sweet olive, 12–20 feet tall and 10–12 feet wide, with large, smooth or finely toothed leaves and usually white flowers in spring. The variety *aurantiacus* has yellow-orange flowers. Native from the Himalayas through southern China to southern Japan. Good small tree. Not for cold-winter climates. USDA: 9–11 Sunset: 8–9, 12–24

O. heterophyllus, hollyleaf osmanthus, 8–15 feet tall and wide, with spiny-edged juvenile leaves, smooth-edged at maturity, and white flowers in fall. Native to central and southern Japan. 'Sasaba', slow growing to 8–10 feet tall and 6 feet wide, has large leaves, deeply cut into sharply pointed, triangular lobes, dark green with prominent, light green veins. 'Goshiki', 4–5 feet tall and wide, has multicolored variegated leaves. May be offered as *O. ilicifolius*. USDA: 7–9 Sunset: 4–10, 14–24

Ozothamnus rosmarinifolius

ROSEMARY EVERLASTING

Evergreen shrub, upright to 4–6 feet tall and wide, with small, linear, almost needlelike, dark green leaves and tight, rounded clusters of tiny white flowers from pinkish red buds in late spring. Native to swampy, open shrublands in southeastern Australia from southern Victoria to Tasmania but grown successfully with moderate to occasional water in moisture-retentive soils. 'Silver Jubilee' has silvery gray leaves. Full sun. May be offered as *Helichrysum*. USDA: 7–9 Sunset: 14–24 WUCOLS: M

^ *Parkinsonia* 'Desert Museum'

Papaver heterophyllum
WIND POPPY

Annual, 1–2 feet tall, with blue-green leaves and fragrant, bright red-orange flowers with crinkled petals in spring. Native to grasslands, chaparral, and open woodlands below 4,000 feet in the Coast Ranges and Sierra Nevada foothills from the San Francisco Bay Area to northern Baja California. Sun or part shade, excellent drainage, infrequent summer water. Needs moisture in growing season to bloom well. May be offered as *Stylomecon*. USDA: N/A Sunset: N/A WUCOLS: N/A

^ *Papaver heterophyllum*

Parkinsonia 'Desert Museum'
PALO VERDE

Deciduous tree or large shrub, fast growing to 20–35 feet tall and wide, with bright green leaves, lime-green bark, and fragrant, funnel-shaped, bright yellow flowers in spring. Nonseeding, thornless hybrid involving three potentially invasive and thorny species from the Southwest and Mexico, *P. aculeata*, *P. florida*, and *P. microphylla*. Drops leaves in summer if grown dry. Full sun, fast drainage, infrequent to no summer water. Best inland; prone to powdery mildew along the coast. May be offered as *Cercidium*. USDA: 8b–11 Sunset: 8–14, 18–20 WUCOLS: L/VL

Parrotia persica
PERSIAN IRONWOOD

Deciduous tree or large shrub, multitrunk, slow growing to 20–35 feet tall and wide, with peeling bark and broad, deeply veined, oval leaves that emerge reddish purple in spring, mature to dark green, and turn yellow, orange, and bright red in fall. Early spring flowers lack petals, but red stamens are showy on bare branches. Native to northern Iran. 'Vanessa' is more upright and columnar. Sun, well-drained soils, moderate to occasional summer water. Best in cool-summer climates. USDA: 4–8 Sunset: 2b–7, 14–17 WUCOLS: L/M

^ *Parrotia persica*

Pelargonium
GERANIUM

Perennials, trailing or upright and shrubby, with aromatic leaves and clusters of showy flowers. Cool sun to part shade or afternoon shade, good drainage, moderate to occasional summer water. Sunset: 8–9, 12–24 WUCOLS: L/M

P. cordifolium, heartleaf geranium, 3–5 feet tall and wide, with heart-shaped green leaves and loose clusters of small, lavender flowers with dark purple veins from midwinter to midsummer. Native to moist places and forest edges in southern coastal areas of Western and Eastern Cape provinces, South Africa. Occasional summer water and part shade. May be offered as *P. cordatum*. USDA: 9b–11

P. ×***hortorum***, zonal geranium, mounding 1–3 feet tall and wide, with medium green, rounded leaves and dense clusters of summer flowers in shades of red, orange, purple, or white. Hybrids of complicated parentage involving *P. zonale* and *P. inquinans*, both native to coastal South Africa. Commonly grown in containers with regular summer water in full sun, but can be grown in the ground with occasional water and some afternoon shade. USDA: 10–11

P. ionidiflorum, celery scented geranium, 1–2 feet tall and spreading by rooting stems, with small, deeply incised, gray-green leaves and purplish pink flowers in summer or almost year-round. Native to inland, semi-arid, summer-rainfall parts of Eastern Cape Province, South Africa. May be offered as 'Pink Fairy Cascades'. USDA: 9–11

P. sidoides, African geranium, 12–18 inches tall and wide, with small, heart-shaped, gray-green leaves and small, dark burgundy, almost black flowers on long, narrow, trailing stems in spring and summer. Native to rocky soils in summer-rainfall grasslands of Eastern Cape Province, South Africa. USDA: 9–11

^ *Pelargonium sidoides*

Penstemon

PENSTEMON

Perennials with green to grayish green, narrowly linear to oval leaves and tubular to bell-shaped flowers in spring, summer, and sometimes fall. The "border" penstemons most often offered in garden centers generally need regular summer water. Those listed here thrive with occasional to infrequent or no summer water in sun or afternoon shade and moisture-retentive soils. All need fast drainage. May be short lived. WUCOLS: L

P. azureus, azure penstemon, 1 foot tall and 2–3 feet wide, with narrow, blue-green leaves and bright blue to lavender flowers. Native to rocky slopes and forest openings between 1,500 and 8,000 feet in the Sierra Nevada foothills and Coast Ranges of northern California and southwestern Oregon. USDA: 7–10 Sunset: N/A

P. centranthifolius, scarlet bugler, 1–2 feet tall and wide, with large, blue-gray to gray-green leaves and bright red to red-orange flowers. Native to dry slopes and seasonally moist meadows below 6,500 feet in the Coast Ranges from northern California to Baja California and the southern Sierra Nevada. Needs especially good drainage. USDA: 5b–10 Sunset: 7–23

^ *Penstemon heterophyllus* 'Catherine de la Mare'

P. heterophyllus, foothill penstemon, 1–2 feet tall and 2–3 feet wide, with glossy, silvery blue-green leaves and showy flowers in an electric combination of pink, purple, and bright blue. Native to coastal mountain ranges of northern and southern California and the northern Sierra Nevada foothills. 'Margarita BOP' and 'Catherine de la Mare' are adaptable and easy garden subjects; both may be hybrids but are usually offered as selections of the species. USDA: 7–10 Sunset: 7–24

P. parryi, Parry's penstemon, 1 foot tall and 2 feet wide, with thick, narrow, bluish green leaves and light to dark pink flowers on stems to 3 feet tall. Native to desert washes and grassy slopes in southern Arizona and northern Mexico. USDA: 8-10 Sunset: 3, 10, 12-13

P. pinifolius, pine leaf penstemon, 1 foot tall and 1-2 feet wide, with needlelike, green leaves and bright red-orange flowers. Native to eastern Arizona, southern New Mexico, and northern Mexico. 'Mersea Yellow' and 'Nearly Red' offer color variations. 'Melon' is 4-6 inches tall with pale orange flowers. USDA: 4-9 Sunset: 1-24

P. spectabilis, showy penstemon, 2-3 feet tall and 3-4 feet wide, with narrow, lance-shaped leaves and showy, purple and blue flowers. Native to dry, rocky slopes below 6,500 feet in coastal mountains of central and southern California and Baja California. Best in hot-summer climates with light shade or afternoon shade. USDA: 7-10 Sunset: 7, 14-23

Peritoma arborea

BLADDERPOD

Evergreen shrub, fast growing to 3-5 feet tall and 4-5 feet wide, with blue-green leaves, clusters of bright yellow flowers almost year-round, and small, pendant, inflated seedpods. Self-sows. Native to a wide range of habitats from coastal bluffs to inland foothills and high or low desert washes in southern California, Baja California, and Arizona. Sun to part shade, fast drainage, infrequent to no summer water. May be offered as *Isomeris*. USDA: 7-10 Sunset: N/A WUCOLS: L/VL

< *Penstemon parryi*

^ *Penstemon pinifolius*

^ *Penstemon spectabilis*

^ *Peritoma arborea*

Perityle incana
GUADALUPE ISLAND ROCK DAISY

Perennial or subshrub, 2–3 feet tall and 3–5 feet wide, with fine-textured, deeply divided, silvery gray-green leaves and clusters of yellow flowers in spring and summer. Native to rocky coastal cliffs and canyons below 4,000 feet on Guadalupe Island off the coast of northern Baja California. Sun to part shade, fast drainage, occasional to infrequent summer water. Cut back in late winter to renew. USDA: 9–10 Sunset: N/A WUCOLS: VL

^ *Perityle incana*

Perovskia atriplicifolia
RUSSIAN SAGE

Perennial, upright to 2–4 feet tall and 2–3 feet wide, with finely dissected, aromatic, gray-green leaves on grayish white stems and a haze of tiny, tubular, lavender-blue flowers on tall, branched spikes in summer. May spread aggressively by rooting stems and can be difficult to control. Native to a wide range of climates and soils in southwestern and central Asia from Turkey and Iran to western China. Sun, good drainage, infrequent summer water. Cut back hard in late winter or early spring to renew. USDA: 4–9 Sunset: 2–24 WUCOLS: L/M

^ *Perovskia atriplicifolia*

Phacelia
PHACELIA

Annuals and perennials with bell-shaped to funnel-shaped spring flowers, usually in one-sided, coiled clusters that uncurl as the flowers open in sequence. Self-sow. Important source of nectar for pollinators, especially bees. Sun or part shade, well-drained soils, infrequent to no summer water. Need moisture in spring to bloom well.

P. bolanderi, woodland phacelia, perennial, mounding, 6 inches tall and 2–3 feet wide, with rough-textured, gray-green leaves and lavender-blue spring or summer flowers in loose, branching clusters. Native to coastal bluffs and canyons of northern California and southern Oregon. Summer dormant if grown dry. USDA: 7–10 Sunset: N/A WUCOLS: M

P. californica, rock phacelia, perennial, 1 foot tall and wide, with velvety, silvery gray-green leaves and purplish pink flowers in coiled clusters on 2-foot stems. Native to coastal bluffs, open slopes, and woodlands in northern California and western Oregon. USDA: 9–10 Sunset: N/A WUCOLS: L/M

P. campanularia, desert bells, annual, 1 foot to 18 inches tall and wide, with oval to rounded, toothed, green to blue-green leaves and bright blue flowers in coiled clusters. Native to open, sandy or rocky places in the deserts of southern California. USDA: N/A Sunset: 1–3, 7–24 WUCOLS: N/A

P. tanacetifolia, lacy phacelia, annual, 1–3 feet tall and wide, with deeply divided, medium green leaves and blue flowers in coiled clusters. Native to sandy or gravelly soils in open woodlands or shrublands below 8,000 feet from the mountains of coastal and inland northern California to the deserts of southern California and east to New Mexico and northern Mexico. USDA: N/A Sunset: 1–3, 7–24 WUCOLS: N/A

^ *Phacelia tanacetifolia*

Philadelphus

MOCK ORANGE

Deciduous and evergreen shrubs with showy clusters of white, usually fragrant flowers with yellow stamens at branch ends in spring to midsummer. Cool sun to light shade, reasonable drainage, moderate to occasional summer water.

P. lewisii, western mock orange, deciduous, 6–10 feet tall and wide, with rough-textured, oval, light green leaves and fragrant flowers in early summer. Native to rocky slopes and woodland openings between 1,000 and 7,000 feet from southwestern British Columbia to northern California and east to Idaho and Montana. CHEYENNE ('PWY01S'), a selection from Wyoming, accepts drier soils. 'Goose Creek', from the Klamath Mountains in California, has double flowers. 'Snow Velvet', from the Cascade foothills in western Oregon, has semi-double flowers. USDA: 4–9 Sunset: 1–10, 14–24 WUCOLS: M

P. mexicanus, Mexican mock orange, evergreen, 6–15 feet tall and wide, with arching branches bearing medium green leaves and intensely fragrant, cup-shaped, white flowers in spring. Native from central Mexico to Guatemala. Moderate summer water. USDA: 9–11 Sunset: 8–9, 14–24 WUCOLS: L/M

P. microphyllus, littleleaf mock orange, deciduous, 4–5 feet tall and wide, with small, oval, bright green leaves and small, fragrant flowers in spring. Native to rocky slopes in desert mountains of southeastern California to Arizona, New Mexico, and northern Mexico. 'Desert Snow' has especially large flowers. USDA: 5–10 Sunset: 1–3, 7, 10, 14–16, 18 WUCOLS: L/M

^ *Philadelphus lewisii*

Philotheca

WAXFLOWER

Evergreen shrubs with aromatic leaves and star-shaped flowers. Native to rocky or sandy soils in dry, open woodlands and shrublands in southeastern Australia. Sun to part shade, most well-drained soils, moderate to occasional or infrequent summer water. May be offered as *Eriostemon*. USDA: 9–10 Sunset: N/A WUCOLS: M

^ *Philotheca myoporoides*

^ *Philotheca salsolifolia*

^ *Phlomis fruticosa*

P. myoporoides, longleaf waxflower, 4–6 feet tall and wide, with dark green, narrowly lance-shaped leaves and small, white flowers from pink-tinged buds in late winter to early spring. 'Profusion' has smaller leaves. 'Galaxy' has especially long flower stems. Best in light shade with moderate to occasional summer water.

P. salsolifolia, 2–6 feet tall and wide, with small, needlelike leaves and pinkish purple flowers in spring and summer. Occasional to infrequent summer water. 'Ballerina', 1–2 feet tall and wide, may be offered as 'Tutu'.

Phlomis

JERUSALEM SAGE

Evergreen shrubs and perennials with felted or woolly, green to gray-green leaves and whorled clusters of tubular flowers at intervals on upright stems in spring and summer. Sun to part shade, fast drainage, good air circulation, occasional to infrequent summer water.

***P.* 'Edward Bowles'**, evergreen shrub, 4–5 feet tall and wide, has especially broad, green leaves and large yellow flowers. Garden origin, believed to be a hybrid between *P. fruticosa* and *P. russeliana*. May be offered as 'Grande Verde'. USDA: 4–10 Sunset: 3b–24 WUCOLS: N/A

P. fruticosa, Jerusalem sage, evergreen shrub, 3–5 feet tall and wide, with gray-green leaves and light yellow flowers. Native to dry, rocky slopes from Crete and Sardinia to western Turkey and Cyprus. USDA: 8b–10 Sunset: 3b–24 WUCOLS: L/M

P. lanata, dwarf Jerusalem sage, evergreen shrub, 1–2 feet tall and 4–6 feet wide, with gray-green leaves and lemon-yellow or yellow-orange flowers. Native to rocky slopes in shrublands and woodland openings from sea level to 5,000 feet on the island of Crete. USDA: 8–11 Sunset: 7–24 WUCOLS: L

P. purpurea, purple phlomis, evergreen shrub, 4–6 feet tall and wide, with gray-green leaves and purplish pink to lavender flowers. Native to rocky slopes from Portugal and Spain to Morocco. The subspecies *caballeroi* has yellow-green leaves. USDA: 9–10 Sunset: 7–24 WUCOLS: L

P. russeliana, Turkish sage, perennial, 1–2 feet tall and 4–5 feet wide, with olive-green leaves and soft yellow flowers on tall stems. Spreads by rhizomes. Native to forest and woodland openings in the mountains of Turkey. USDA: 7–10 Sunset: 2–24 WUCOLS: L/M

^ *Phlomis purpurea*

Phormium

NEW ZEALAND FLAX

Perennials with upright or arching, strap-shaped leaves and tubular flowers on branched stalks in late spring or early summer. Sun to part shade, good drainage, moderate to occasional or infrequent summer water. USDA: 7b–11 Sunset: 7–9, 14–24

P. cookianum, mountain flax, 4–5 feet tall and 6–8 feet wide, with arching, olive-green leaves and greenish yellow flowers on arching stalks. Native to New Zealand, both in the mountains and in exposed coastal areas. 'Tricolor', 3 feet tall, has dark green leaves with creamy white margins. 'Cream Delight', a sport of 'Tricolor', has creamy yellow leaves with green margins edged in red. Best along the coast or in light shade or afternoon shade inland. May be offered as *P. colensoi*. WUCOLS: M

P. tenax, coastal flax, upright to 8–10 feet tall and 6–8 feet wide, with bronzy green leaves and red-orange flowers on upright stalks. Native to various habitats in New Zealand, from seasonally wet places and along streams to coastal cliffs. 'Dusky Chief', 6–8 feet tall, has variable leaves, pale to dark maroon when new, silvery gray beneath, and older leaves may be a smoky purplish gray. 'Shiraz', 3 feet tall, has dark burgundy red leaves. 'Tom Thumb', 2–3 feet tall and wide, has wavy edged olive-green leaves with dark bronzy red margins. WUCOLS: L/M

^ *Phormium tenax* 'Dusky Chief'

^ *Phylica pubescens*

^ *Pinus attenuata*

Most plants available are hybrids between *P. tenax* and *P. cookianum*. Some may grow much larger than expected, and those with variegated or unusually colored leaves may revert over time to solid bronze or green. Among the small to medium-sized hybrids that tend to retain their colors are *P.* 'Apricot Queen', pale yellow leaves with green margins; *P.* 'Bronze Baby', reddish bronze leaves; *P.* 'Dark Delight', especially dark reddish brown leaves; *P.* 'Dazzler', dark maroon and bright red leaves; and *P.* 'Jack Spratt', twisted dark purple leaves.

Phylica pubescens
FEATHERHEAD

Evergreen shrub, upright to 4–6 feet tall and wide, with narrow, needlelike, silvery gray-green leaves covered in white or yellowish hairs, and small creamy white fall to late-winter flowers hidden among feathery, creamy white bracts. Native to dry sandstone cliffs east of Cape Town, South Africa. Sun, excellent drainage, good air circulation, occasional to infrequent or no summer water. May be offered as *P. plumosa*, which is a smaller plant. USDA: 9–10 Sunset: N/A WUCOLS: L/M

Pinus
PINE

Evergreen coniferous trees with needlelike leaves. Many are too large at maturity for most gardens and some need regular summer water. Those listed here are smaller trees or shrubs that thrive with occasional to infrequent or no summer water in sun to part shade. All need good to excellent drainage.

P. attenuata, knobcone pine, fast growing to 25–40 feet tall and 20–25 feet wide, often multitrunk, with long, green to yellowish green needles on horizontal branches that turn up at the ends. Native to dry, gravelly or sandy soils from sea level to 5,500 feet in the mountains of southwestern Oregon and northern California and intermittently south along the coast to Baja California. USDA: 7–9 Sunset: 2–10, 14–21 WUCOLS: L/VL

P. contorta, shore or lodgepole pine, highly variable, with three or sometimes four forms recognized as native from southeastern Alaska to northern Baja California and from coastal to high montane habitats. Those from higher elevations can become 100-foot trees. Coastal forms, 20–35 feet tall and wide, are more often available and better suited to residential gardens. Some cultivars are quite small. 'Spaan's Dwarf' is 3 feet tall and 4 feet wide with dark green needles and an irregular, open habit. 'Taylor's Sunburst', 6–8 feet tall and 2–3 feet wide, has bright yellow new spring growth that matures to yellowish green. Best along the coast or in cool-summer climates, where most need moderate to occasional summer water. Not low water in hot-summer climates. USDA: 7–10 Sunset: 4–9, 14–24, A3 WUCOLS: M

^ *Pinus contorta*

P. edulis, pinyon or Colorado pinyon pine, slow growing to 20–50 feet tall and 10–25 feet wide, with short, blue-green needles that mature to yellowish green and an irregular, spreading habit. Native to dry, rocky slopes and mesas between 3,000 and 8,000 feet from Utah and Colorado to Arizona and New Mexico. May be offered as a variety of *P. cembroides* or of *P. monophylla*. Needs occasional summer water. USDA: 6–8 Sunset: 1–11, 14–21 WUCOLS: L/VL

P. eldarica, Afghan pine, fast growing to 30–50 feet tall and 25 feet wide, with long, dark green needles and an upright, symmetrical habit. Native to dry, rocky soils on north-facing slopes in the Caucasus Mountains of northern Azerbaijan near the border with Georgia. May be offered as a subspecies of *P. brutia*, a similar species native to the northeastern Mediterranean region. *P. halepensis*, from the western Mediterranean, is also similar. USDA: 6–10 Sunset: 6–9, 11–24 WUCOLS: N/A

P. monophylla, single-leaf pinyon, slow growing to 15–30 feet tall and wide, pyramidal when young, irregular and more open at maturity, with stiff, gray-green or blue-gray needles on upswept branches. Native to dry, rocky slopes and ridges from 3,000 to 10,000 feet in the mountains of southern California to Arizona, New Mexico, and northern Baja California. USDA: 5–10 Sunset: 2–12, 14–21 WUCOLS: L/VL

P. mugo, mugo pine, variable, usually multitrunk, dense and low branching, 15–20 feet tall and 20–25 feet wide, with dark green to dark grayish green needles. Native to mountains of central and southern Europe from Spain to the Balkan Peninsula. 'Slowmound', a selection from Oregon, is a little more than 1 foot tall and 2 feet wide. Best in cool-summer climates with occasional summer water. USDA: 3–7 Sunset: 1–11, 14–24 WUCOLS: L/M

P. torreyana, Torrey pine, fast growing to 25–50 feet tall and 30–40 feet wide, much larger with summer moisture, with long, gray-green needles and an open, irregular habit. Native to coastal southern California and the Channel Islands. Best along the coast in sandy or rocky soils. USDA: 8–10 Sunset: N/A WUCOLS: L/M

^ *Pistacia chinensis*

Pistacia
PISTACHE

Deciduous, semi-evergreen, or evergreen trees with compound leaves divided into narrow leaflets and clusters of inconspicuous greenish flowers. Sun, most well-drained soils, occasional to infrequent or no summer water.

P. chinensis, Chinese pistache, deciduous, 30–40 feet tall and wide, with leaves that turn brilliant red, orange, or yellow in fall. Native to summer-rainfall climates of central and southeastern China, Taiwan, and the Philippines. Fruit of female plants can be messy and plants may spread by seed. 'Keith Davey' is a fruitless male selection. Best near the coast or in cool-summer climates. Considered invasive in some parts of California. USDA: 6–9 Sunset: 4–23 WUCOLS: L/M

P. lentiscus, mastic, evergreen, 10–30 feet tall and wide, with leathery, dark green leaves. Native to a variety of habitats in the Mediterranean region, mostly near the coast, from Spain and Morocco to Turkey and Israel. Good hedge, screen, or small multitrunk tree. USDA: 9–11 Sunset: 8–9, 12–24 WUCOLS: VL/M

Polypodium californicum
CALIFORNIA POLYPODY

Fern, 12–18 inches tall and slowly spreading by rhizomes, with bright green fronds arising at intervals. Native to canyons, streambanks, and north-facing slopes along the coast and in coastal mountains from northern California to northern Baja California. Full shade to filtered sun, most soils, infrequent to no summer water. Deciduous and dormant in summer if grown dry, reviving with fall rains. USDA: 7–10 Sunset: 4–9, 14–17, 22–24 WUCOLS: VL

^ *Polypodium californicum*

Polystichum munitum
WESTERN SWORD FERN

Fern, 2–3 feet tall and wide, upright and spreading, with leathery, dark green fronds. Larger in humid, cool-summer parts of the Pacific Northwest. Native to wooded slopes from sea level to 3,000 feet and from southeastern Alaska to Guadalupe Island off the coast of Baja California. Full shade to filtered sun, well-drained, humusy soils, moderate to occasional or infrequent summer water. Best near the coast. USDA: 5–9 Sunset: 2–9, 14–24, A3 WUCOLS: M/H

^ *Polystichum munitum*

Prunus
CHERRY

Evergreen and deciduous shrubs and small trees with glossy, dark green leaves, clusters of small, creamy white flowers in spring, and red to dark purple fruits in fall. Sun to part shade, most well-drained soils, occasional to infrequent or no summer water. Fruit drop can be messy.

P. ilicifolia, hollyleaf cherry, evergreen, 15-25 feet tall and 10-15 feet wide, with spiny-edged leaves. Native to coastal mountains and foothills in central and southern California, the Channel Islands, and Baja California, mostly on west-facing slopes. Subspecies *lyonii*, Catalina cherry, is similar but with smooth-edged leaves and longer flower clusters. Native to the Channel Islands in southern California and the Baja California mainland. May be offered as *P. lyonii*. USDA: 8-11 Sunset: 5-9, 12-24 WUCOLS: L/VL

P. virginiana* var. *demissa, western chokecherry, deciduous, upright to 15-20 feet tall and wide, with smooth-edged leaves. Fast growing and short lived. Native to many habitats, mostly between 6,000 and 10,000 feet and often near springs or seasonal streams, from British Columbia south to northern Mexico and Texas. Best with afternoon shade in hot-summer climates. USDA: 5-10 Sunset: 2-10, 14-24 WUCOLS: N/A

^ *Prunus ilicifolia* (behind ceanothus shrubs)

Puya

PUYA

Succulent or semi-succulent rosettes of long, narrow leaves with aggressively spiny margins and waxy, trumpet-shaped, blue-green flowers. Spread slowly by offsets to form a large colony. Native to dry, rocky slopes in the Andes of Chile and Argentina. Sun to light shade, excellent drainage, occasional summer water. Need dry soil in winter. Best in southern California. USDA: 9-11 WUCOLS: L/VL

^ *Puya berteroniana*

P. alpestris, sapphire tower, 2-3 feet tall and wide, with green to gray-green leaves and blue-green flowers on 3- to 4-foot stalks. Sunset: N/A

P. berteroniana, turquoise puya, 3-4 feet tall and wide, with gray-green leaves and metallic turquoise blue flowers on 6- to 8-foot stalks. May be offered as *P. berteroana*. Sunset: 9, 13-17, 19-24

P. coerulea* var. *coerulea, silver puya, 2-3 feet tall and wide, with silvery white leaves and dark blue-violet flowers from pink buds on pink stems. May be offered as *P. coerulea* var. *violacea*, a similar plant with silvery green leaves. Sunset: N/A

^ *Puya coerulea* var. *coerulea*

Quercus

OAK

Evergreen and deciduous trees and shrubs, some too large for most residential gardens. Oaks of the summer-dry Pacific coast are found in many habitats, coastal and inland, and from low to high elevations. Sun to part shade, most well-drained soils, occasional to infrequent or no summer water.

Q. agrifolia, coast live oak, evergreen, 35–75 feet tall and 35–50 feet wide, dense and broadly rounded, with leathery, dark green, spiny-margined leaves. Native from northern to southern California and northern Baja California along the coast and in the coastal mountains, usually below 2,500 feet. No summer water. Best in mild climates. Young trees do best in part shade. USDA: 8–10 Sunset: 7–9, 14–24 WUCOLS: L/VL

Q. chrysolepis, canyon live oak, evergreen, variable, 30–60 feet tall and wide, rounded and wide-spreading, with leathery, dark green leaves with smooth or toothed edges. Native to a wide range of elevations, soils, and moisture regimes from southwestern Oregon to Baja California and east into Nevada and Arizona. Occasional to infrequent or no summer water. USDA: 6–9 Sunset: 3–11, 14–24 WUCOLS: L/VL

Q. douglasii, blue oak, deciduous, slow growing to 30–60 feet tall and wide, with dark bluish green, slightly lobed and wavy-edged leaves that turn red and yellow in fall. Native to dry, rocky slopes below 3,500 feet in foothills surrounding California's Central Valley north to the Klamath Mountains and south to the Tehachapi Mountains. Infrequent to no summer water. Drops some leaves in summer if grown dry. USDA: 6–9 Sunset: 3–11, 14–24 WUCOLS: L/VL

Q. engelmannii, Engelmann or mesa oak, evergreen, variable, slow growing to 40–50 feet tall and 60–80 feet wide, with bluish gray-green, shallowly

^ *Quercus agrifolia*

^ *Quercus engelmannii*

lobed, usually smooth-edged leaves. Native to grasslands and woodlands, often near streams or springs, in coastal mountains of southern California and northwestern Baja California. Occasional summer water if winter rains fail. Drops some leaves in summer if grown dry. USDA: 8–11 Sunset: 7–9, 14–24 WUCOLS: L/VL

Q. garryana, Oregon white or Garry oak, deciduous, 30–60 feet tall and 30–50 feet wide, or multitrunk and shrubby to 20 feet tall, with large, deeply lobed, glossy, dark green leaves and furrowed, pale whitish gray bark. Native to both dry, rocky slopes and seasonally flooded riparian areas, mostly west of the Cascades, from southwestern British Columbia to the Tehachapi Mountains north of Los Angeles. Infrequent to no summer water. USDA: 6–9 Sunset: 2a–11, 14–23 WUCOLS: L

Q. hypoleucoides, silverleaf oak, evergreen, broadly rounded to 30-50 feet tall and 20-40 feet wide, with lance-shaped, leathery, gray-green leaves, felted and silvery white beneath. Native to the mountains of nothern Mexico and southeastern Arizona to Texas at mid- to high elevations. Occasional to infrequent summer water. USDA: 7-10 Sunset: N/A WUCOLS: N/A

Q. tomentella, island oak, evergreen, 30-50 feet tall and 20-30 feet wide, with leathery, elliptical, dark green leaves with widely spaced teeth, hairy and paler beneath. Native to canyons and ridges below 2,100 feet on the Channel Islands in southern California and Guadalupe Island in Baja California. Best right along the coast. Infrequent to no summer water. May be offered as a subspecies of *Q. chrysolepis*. USDA: 7-10 Sunset: 7-9, 14-17, 19-24 WUCOLS: L

Q. wislizeni, interior live oak, evergreen, variable, to 30-70 feet tall and wide, with glossy, green leaves with smooth or spiny margins. Native to dry valleys and foothills in the Coast Ranges and Sierra Nevada from southern Oregon to Baja California and from sea level to 5,000 feet. The variety *frutescens* is a shrubby form about 15-25 feet tall. Infrequent to no summer water. USDA: 8-10 Sunset: 7-9, 14-16, 18-21 WUCOLS: VL/M

Ranunculus californicus

CALIFORNIA BUTTERCUP

Perennial, 1-2 feet tall and wide, with bright green, lobed and divided basal leaves and glossy, clear yellow flowers on leafless stems in spring. Dormant and deciduous in summer. May self-sow. Native to seasonally moist sites below 7,000 feet, both along the coast and inland, from southern Oregon to northwestern Baja California; also found on Vancouver Island and coastal mainland British Columbia. Sun to part shade, most soils, no summer water. Other species of *Ranunculus* need regular water. USDA: 8-11 Sunset: N/A WUCOLS: VL

^ *Ranunculus californicus*

^ *Rhagodia spinescens*

Rhagodia spinescens

AUSTRALIAN SALTBUSH

Evergreen shrub, 1-3 feet tall and 6-8 feet wide, with silvery gray-green leaves, inconspicuous flowers, and showy clusters of bright red berries in fall. Native to central and eastern Australia, both along the coast and inland. Sun to light shade, most well-drained soils, infrequent summer water. Good seaside plant. USDA: 9-11 Sunset: N/A WUCOLS: L/VL

Rhamnus

BUCKTHORN

Evergreen shrubs and small trees with oval to lance-shaped, usually dark green leaves, inconspicuous flowers, and showy, berrylike fruit. Sun to part shade, most well-drained soils, occasional to infrequent or no summer water.

R. alaternus, Italian buckthorn, fast growing to 12–15 feet tall and 6–8 feet wide, with glossy, dark green leaves and red fruit that ages to dark purple. Native to the Mediterranean region from Portugal to Greece and northeast to the western shores of the Black Sea. 'Variegata', 8–12 feet tall, has dark green leaves edged in creamy white. Considered a high risk for invasiveness in parts of California. USDA: 7–9 Sunset: 4–24 WUCOLS: L/M

R. californica*, see *Frangula

R. crocea, spiny redberry, 3–6 feet tall and wide, with small, spiny, dark green leaves and shiny, bright red fruit. Native to Coast Ranges below 3,000 feet in central to southern California, Baja California, and east to Arizona. Afternoon shade inland. USDA: 7–10 Sunset: 7, 14–24 WUCOLS: L/VL

R. ilicifolia, hollyleaf redberry, slow growing to 8–10 feet tall and wide, with thick, glossy, green leaves with spiny margins and bright red fruit. Native to rocky slopes and canyons from 500 to 6,500 feet in mountains and foothills of southern Oregon, California, Arizona, and Baja California. May be offered as a subspecies of *R. crocea*. USDA: 7–10 Sunset: 7–10, 14–23 WUCOLS: L/VL

^ *Rhamnus alaternus* 'Variegata'

^ *Rhaphiolepis indica* 'Ballerina'

Rhaphiolepis

INDIA HAWTHORN

Evergreen shrubs with leathery, dark green leaves, new growth often coppery, and clusters of small, star-shaped, lightly fragrant, pink or white flowers in spring. Sun to light shade, most well-drained soils, good air circulation, moderate to occasional or infrequent summer water. Good coastal or inland in California but may develop foliar disease in wetter parts of the Pacific Northwest. USDA: 8–10 Sunset: 8–10, 12–24 WUCOLS: L/M

R. indica, India hawthorn, 4–5 feet tall and 5–6 feet wide, with pink-tinged, white flowers. Native to southern China south to Cambodia and Vietnam. 'Ballerina' is 2–3 feet tall and 4 feet wide with deep pink flowers. 'Clara' is 3–5 feet tall and wide with white flowers.

R. MAJESTIC BEAUTY **('Montic')**, 12–20 feet tall and 8–10 feet wide, has large leaves and large clusters of light pink flowers. Usually trained as a small tree. Hybrid of uncertain parentage, possibly a cross between *R. umbellata* and an *Eriobotrya* species.

R. umbellata, Yeddo hawthorn, slow growing to 6–8 feet tall and wide, with white flowers. Leaves are smaller and darker green than those of *R. indica* and branches are more upright. Native to Japan, Korea, Taiwan, and coastal mainland China. 'Minor' is only slightly smaller, 4–6 feet tall and 3–4 feet wide. More resistant to foliar disease than *R. indica*.

Rhodanthemum hosmariense

MOROCCAN DAISY

Perennial, less than 1 foot tall and 1–2 feet wide, with finely divided, silvery green leaves and pure white flowers with bright yellow centers in spring and again in fall. Native to rocky, exposed sites in the Atlas Mountains of Morocco and Algeria. Sun, excellent drainage, occasional to infrequent summer water. May be offered as *Chrysanthemum*, *Leucanthemum*, or *Pyrethropsis*. USDA: 8–10 Sunset: 14–24 WUCOLS: L/M

^ *Rhodanthemum hosmariense*

Rhus

SUMAC

Evergreen and deciduous shrubs with dense clusters of tiny flowers and red, berrylike fruits. Those listed here are evergreen. Sun to light shade, most well-drained soils, infrequent to no summer water. USDA: 8–10 WUCOLS: L/VL

R. integrifolia, lemonade berry, 3–12 feet tall, lower along the coast than inland, and 10–15 feet wide. Leathery, dark green leaves, usually with small, sharp marginal teeth and pink or white flowers in midwinter to spring. Native to dry, open slopes below 3,000 feet in coastal central and southern California from Santa Barbara County and the Channel Islands to northern Baja California. Sunset: 8–9, 14–17, 19–24

^ *Rhus integrifolia*

^ *Rhus ovata*

R. ovata, sugar bush, 6–15 feet tall and wide, similar to *R. integrifolia* but with larger, darker green leaves and adapted to inland rather than coastal locations. Native to dry slopes and canyons below 4,000 feet in foothills and mountains of southern California, Arizona, and northern Baja California. Sunset: 9–12, 14–24

Ribes

FLOWERING CURRANT, GOOSEBERRY

Deciduous and evergreen shrubs, spreading by rhizomes, with lobed, sometimes aromatic, maple-like leaves and small, sometimes fragrant, tubular flowers followed by smooth or bristly berries. Gooseberries have sharp spines and flowers displayed singly along the stems. Currants have no spines and flowers in pendant clusters. Some need regular water and shade. Those listed here thrive with moderate to occasional or no summer water in cool sun or part shade and well-drained soils.

R. aureum, golden currant, deciduous, 6–8 feet tall and wide, with light green leaves that turn reddish purple in fall and fragrant, bright yellow flowers. Native to dry slopes and moist streambanks in much of North America. The variety *gracillimum*, native to foothills of the Transverse Ranges in southern California, is 3–6 feet tall and sprawling. Moderate to occasional summer water. Accepts periodic flooding. USDA: 5–10 Sunset: 1–12, 14–23, A2–3 WUCOLS: L/VL

R. malvaceum, chaparral currant, deciduous, 5–8 feet tall and 5–6 feet wide, with gray-green, aromatic leaves that turn red in fall and fragrant pink flowers. Drops leaves in summer if grown dry. Native to woodland, forest, and chaparral from the San Francisco Bay Area and inner North Coast Ranges to northern Baja California. 'Dancing Tassels', a selection from San Clemente Island off the coast of southern California, has especially long clusters of white and pale rose flowers. *R. malvaceum* var. *viridifolium* 'Ortega Beauty' has deep rose, almost red, flowers. Infrequent to no summer water. USDA: 7–10 Sunset: 6–9, 14–24 WUCOLS: L/VL

R. sanguineum, red-flowering currant, deciduous, 6–10 feet tall and wide, with medium green leaves and reddish pink flowers. Native along the coast and in the Coast Ranges from southern British Columbia to southern California and east to Idaho.

^ *Ribes sanguineum*

^ *Ribes sanguineum* var. *glutinosum* 'Inverness White'

R. sanguineum var. *sanguineum* 'King Edward VII', a nineteenth-century selection from England, has red flowers and dark green to bluish green leaves. *Ribes sanguineum* var. *glutinosum* 'Inverness White' is a white-flowered selection from Marin County, California. 'Oregon Snowflake', a compact plant to 4 feet tall with deeply dissected leaves, also has white flowers. Moderate to occasional or infrequent summer water. USDA: 6–10 Sunset: 4–9, 14–24, A3 WUCOLS: L/M

R. speciosum, fuchsia-flowered gooseberry, semi-evergreen, 6–8 feet tall and wide, with bright green leaves on sharply spiny stems and red flowers with protruding red stamens. Drops leaves in summer if grown dry. Native to shady, north-facing slopes or seasonally moist places along the coast from the San Francisco Bay Area to northern Baja California. 'Rana Creek' is an especially free-flowering selection from Carmel Valley in coastal central California. Occasional to infrequent or no summer water. USDA: 7–10 Sunset: 7–9, 14–24 WUCOLS: VL/L/M

R. viburnifolium, Catalina currant, evergreen, 2–3 feet tall and 6–10 feet wide, with glossy, dark green leaves on red stems and red flowers. Native to Santa Catalina Island and the coastal mainland of southern California and northern Baja California. 'Spooner's Mesa', a selection from southern San Diego County, has larger leaves and a denser habit. Occasional to infrequent summer water. Good for dryish shade. USDA: 8–10 Sunset: 5, 7–9, 14–17, 19–24 WUCOLS: VL/L/M

^ *Ribes speciosum*

^ *Romneya coulteri*

Romneya coulteri

MATILIJA POPPY

Perennial from rhizomes, upright to 6–10 feet tall and 2–4 feet wide, with bluish gray stems and leaves and large flowers with crinkly white petals and bright yellow stamens in early summer. Spreads vigorously once established. Native to coastal sage scrub and chaparral in dry canyons below 4,000 feet in southern California and Baja California. 'White Cloud' has especially large flowers. Dormant in late summer if grown dry. Full sun, excellent drainage, infrequent to no summer water. Can be cut back hard in late summer or fall. USDA: 6–11 Sunset: 4–12, 14–24 WUCOLS: L/VL

Rosa

ROSE

Deciduous and evergreen, usually thorny or prickly shrubs and vines with showy flowers in many different colors. Most roses offered are modern cultivars that prefer moderate to regular summer water, although many can be grown successfully with less. Those listed here are content with occasional to infrequent or no summer water in moisture-retentive soils. Full sun or afternoon shade and good drainage. Mulch to maintain moisture and keep roots cool.

R. banksiae, Lady Banks' rose, evergreen in mild climates, climber to 20 feet or more, with glossy, green leaves on stems without thorns and masses of small flowers in early to midspring. Native to central and western China. 'Lutea' has unscented,

^ *Rosa* 'Mutabilis'

double yellow flowers. *R. banksiae* var. *banksiae* (sometimes offered as *R. banksiae* 'Alba Plena') has scented, double white flowers. Occasional summer water. USDA: 7–10 Sunset: 4–24 WUCOLS: L/M

R. glauca, red-leaf rose, deciduous, 5–8 feet tall and wide, with purplish gray-green leaves on deep purple stems with few thorns and clusters of single pink flowers. Native to the mountains of central and southern Europe. Occasional summer water. May be offered as *R. rubrifolia*. USDA: 4–9 Sunset: 1–24, A1–3 WUCOLS: N/A

R. minutifolia, Baja rose, evergreen, slow growing to 2–3 feet tall and wide, with glossy, green leaves divided into tiny, toothed leaflets. Intensely thorny branches and pink, purplish pink, or sometimes white flowers in winter and early spring. Slowly thicket forming. Native to coastal bluffs and foothills in the summer fog zone of northern Baja California and southwestern San Diego County. Drops leaves in summer if grown dry. Best in coastal southern California. Infrequent to no summer water. USDA: 9–10 Sunset: N/A WUCOLS: L/VL

R. 'Mutabilis', China rose, evergreen, 6–8 feet tall and wide, with glossy, dark green leaves, new growth bronzy, thorny branches, and creamy yellow flowers that age to deep rose-pink. Origin obscure; possibly a garden hybrid from Italy. May be offered as *R. ×odorata* 'Mutabilis' or as *R. chinensis* 'Mutabilis'. Occasional summer water. USDA: 6–10 Sunset: 4–9, 12–24 WUCOLS: N/A

^ *Rosmarinus officinalis*

Rosmarinus officinalis

ROSEMARY

Evergreen shrub, upright to sprawling or prostrate, with aromatic, needlelike, dark green leaves and small blue flowers in late winter through spring and often again in fall. Native to coastal cliffs and islands of the Mediterranean region from Spain to Greece and around the northern edge of the Black Sea. Sun, well-drained soils, occasional to infrequent or no summer water. Tip prune as needed to maintain form. May be offered as *Salvia rosmarinus*. USDA: 9–11 Sunset: 4–24 WUCOLS: VL/L/M

Upright *R. officinalis* selections, 5–8 feet tall, include 'Blue Spires', 'Spice Islands', and 'Tuscan Blue', all with dark blue flowers. Midheight, semi-upright spreaders good for bank covers are 'Collingwood Ingram', 'Ken Taylor', and 'Mozart', all with deep blue flowers. Nearly flat spillers, good for draping down walls, include 'Huntington Carpet', with pale blue flowers, IRENE ('Renzels'), with lavender-blue flowers, and Prostratus Group, with pale lavender-blue flowers. 'Arp' and 'Madalene Hill' ('Hill Hardy', 'Madeline Hill'), both upright to 3 feet tall, are hardy to USDA zone 7.

^ *Rosmarinus officinalis* 'Tuscan Blue'

^ *Rosmarinus officinalis* Prostratus Group

^ *Rubus rolfei*

Rubus rolfei
TAIWAN BRAMBLE

Evergreen shrub, to 1 foot tall and spreading to 5 feet wide by rooting stems, with small, rough-textured, dark green, rounded and lobed leaves that turn red in fall, small white flowers, and reddish yellow fruit. Cascades over walls. Native to mountains of Taiwan. 'Emerald Carpet', a selection from Vancouver, British Columbia, has larger leaves. Cool sun to shade, most well-drained soils, infrequent to no summer water. Best in cool-summer climates. May be offered as *R. pentalobus*, *R. calycinoides*, or *R. hayata-koidzumii*. USDA: 7–9 Sunset: 4–6, 14–17 WUCOLS: M

^ *Salvia* 'Bee's Bliss'

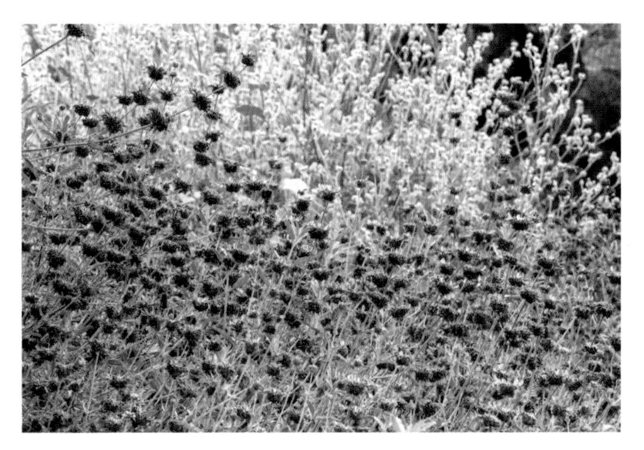

^ *Salvia clevelandii*

Salvia
SAGE

Evergreen and deciduous shrubs and perennials with green to gray-green, aromatic leaves and tubular flowers in spring or summer. Many sages need regular water. Those listed here thrive with moderate to no summer water with fast drainage and good air circulation. Many accept or prefer part-day shade inland.

S. 'Allen Chickering', evergreen shrub, 3–5 feet tall and wide, with large, bright lavender-blue flowers. *S.* 'Aromas' and *S.* 'Pozo Blue' are similar and all are hybrids between *S. clevelandii* and *S. leucophylla*. Infrequent summer water. USDA: 8–10 Sunset: 8–9, 12–24 WUCOLS: L/VL

S. apiana, white sage, evergreen shrub, slow growing to 3–5 feet tall and 4–6 feet wide, with large, velvety, gray-green leaves that age to silvery white and tall spikes of lavender-tinged, white flowers. Native to dry, gravelly, exposed hillsides below 5,000 feet from Santa Barbara County to northwestern Baja California. Infrequent to no summer water. USDA: 9–11 Sunset: 7–9, 11, 13–24 WUCOLS: L/VL

S. 'Bee's Bliss', evergreen shrub, 1–2 feet tall and 6–8 feet wide, with dense, gray-green to gray-white leaves and lavender-blue flowers. Hybrid of garden origin, likely between *S. leucophylla* and *S. sonomensis* or *S. clevelandii*. Infrequent summer water. USDA: 8–10 Sunset: 7–9, 14–24 WUCOLS: L

S. clevelandii, Cleveland sage, evergreen shrub, 3–5 feet tall and 4–6 feet wide, with felted, gray-green, rough-textured leaves and blue-violet flowers. Native to dry chaparral and coastal sage scrub below 4,500 feet in southern California and northwestern Baja California. 'Winnifred Gilman' is more compact and has especially rich violet flowers. Infrequent summer water. USDA: 8–11 Sunset: 8–9, 12–24 WUCOLS: L/VL

S. 'Dara's Choice', evergreen shrub, fast growing to 1–2 feet tall and 3–5 feet wide, with grayish green leaves and blue-violet flowers. Hybrid of garden origin, likely between *S. sonomensis* and *S. mellifera*. Infrequent summer water. USDA: 8–10 Sunset: 7, 14–24 WUCOLS: L

S. darcyi, Darcy sage, perennial, 2–4 feet tall and 4–6 feet wide, with pale green, triangular leaves and bright red flowers over a long season. Native to rocky, limestone soils at 9,000 feet in the mountains of northeastern Mexico. Deciduous in winter. Moderate summer water. USDA: 7–10 Sunset: 8–9, 12, 14–24 WUCOLS: L/M

S. 'Desperado', evergreen shrub, 6–8 feet tall and wide, with whitish gray leaves and lavender-pink flowers. Hybrid of *S. apiana* and *S. leucophylla*. Infrequent to no summer water. USDA: 8–10 Sunset: N/A WUCOLS: N/A

S. greggii, autumn sage, evergreen or deciduous shrub, 2–3 feet tall and wide, with narrow, light green leaves and bright red flowers. Native to rocky slopes at 5,000 to 9,000 feet in southwestern Texas and northeastern Mexico. Many cultivars with red, pink, orange, purple, or white flowers. Deciduous in cold winters. Afternoon shade in hot-summer climates. Occasional summer water. USDA: 8–10 Sunset: 8–24 WUCOLS: L/M

S. ×jamensis, Jame sage, evergreen shrub, 2–3 feet tall and wide, with glossy green leaves and flowers in many colors. Natural hybrids found in northeastern Mexico near the town of Jame, where the parents, *S. greggii* and *S. microphylla*, share territory. Many cultivars. Moderate to occasional summer water. Best in mild-winter, mild-summer climates. USDA: 8–10 Sunset: 8–24 WUCOLS: L/M

S. lanceolata, lanceleaf or Rocky Mountain sage, perennial, 3–4 feet tall and wide, with silvery gray-green leaves and bicolored, grayish rose and yellow flowers. Native to sandy or rocky soils along the west coast of South Africa from the Cape Peninsula to Namaqualand. Moderate to occasional summer water. USDA: 9–11 Sunset: N/A WUCOLS: L

S. leucantha, Mexican bush sage, evergreen shrub, fast growing to 3–4 feet tall and 4–6 feet wide, with gray-green, lance-shaped leaves and tall spikes of tiny white flowers with purple bracts. Native to tropical and subtropical forests in central and eastern Mexico. 'Santa Barbara' is 30 inches tall and 4 feet wide with purple flowers. Occasional to infrequent summer water. Cut back in fall or late winter to renew. USDA: 8–10 Sunset: 12–24 WUCOLS: L

^ *Salvia darcyi*

^ *Salvia* 'Desperado'

^ *Salvia lanceolata*

S. leucophylla, purple sage, evergreen shrub, 4-6 feet tall and wide, with wrinkled, whitish gray, lance-shaped leaves, new growth green, and rosy lavender flowers. Native to gravelly soils on hot, dry hillsides below 4,000 feet in coastal southern California and Baja California. 'Figueroa' has exceptionally white leaves. 'Point Sal' is 2-3 feet tall and 8-10 feet wide. Infrequent to no summer water. USDA: 8-10 Sunset: 8-9, 14-17, 19-24 WUCOLS: VL/M

S. rosmarinus, see Rosmarinus officinalis

S. spathacea, hummingbird sage, perennial, 1-3 feet tall and spreading by rhizomes to 4-5 feet wide, with long, bright green, wrinkled leaves and pale pink to dark rose-colored flowers. Native to lightly shaded slopes below 2,000 feet in the Coast Ranges from central to southern California. Good choice for dryish shade. Occasional to infrequent summer water. USDA: 8-11 Sunset: 7-9, 14-24 WUCOLS: L/M

^ *Salvia leucantha*

^ *Salvia leucophylla* 'Point Sal'

Sambucus cerulea

WESTERN BLUE ELDERBERRY

Deciduous shrub or small tree, fast growing to 10-25 feet tall and wide, with dark green leaves divided into serrated leaflets and large, flat-topped clusters of midsummer white flowers followed by dark blue, berrylike fruit. Native to moist places within dry, open areas from southern British Columbia and western Alberta south to California, northwestern Mexico, and western Texas. May be offered as *S. caerulea* or as a subspecies of *S. nigra* or of *S. mexicana*. Cool sun to light shade, well-drained soils, moderate to occasional summer water. Not low water in hot-summer climates or hot locations. USDA: 5-10 Sunset: 2-24 WUCOLS: L/M

^ *Sambucus cerulea*

Santolina
SANTOLINA

Evergreen shrubs with aromatic, dissected leaves and small, rounded clusters of tiny flowers in summer. Full sun, excellent drainage, infrequent summer water. Cut back in late winter or early spring to renew. Plants may need replacing every few years. USDA: 6–10 WUCOLS: L

S. chamaecyparissus, gray santolina, 1–2 feet tall and 2–3 feet wide, with silvery gray leaves and bright yellow flowers. Native to dry, rocky slopes from Portugal to Italy and Croatia. Sunset: 2–24

S. rosmarinifolia, green santolina, 1–2 feet tall and wide, with bright green leaves and creamy yellow flowers. Native to Portugal, Spain, and southern France. May be offered as *S. virens* or *S. viridis*. Sunset: 3–9, 14–24

^ *Santolina chamaecyparissus*

^ *Santolina rosmarinifolia*

Schizachyrium scoparium
LITTLE BLUESTEM

Warm-season bunchgrass, upright or densely mounding to 1–3 feet tall and wide, with narrow leaves, intensely blue in spring, green to bluish green in summer, and bronzy green, orange, or coppery red in fall. Fluffy, silvery flowers on pink, 4- to 5-foot stems in late summer. Self-sows but not aggressively. Native to prairies and open woodlands in much of North America. 'Blaze' turns deep red in fall and pink in winter. 'The Blues' has light blue leaves that turn purplish in fall and winter. Sun, most soils, moderate to occasional summer water. Cut back in late winter or early spring to renew. USDA: 3–9 Sunset: 1–24 WUCOLS: L

^ *Schizachyrium scoparium*

^ *Sedum cauticola*

^ *Sedum palmeri*

Sedum

STONECROP

Succulents, low mats to upright, with fleshy stems and leaves in a wide range of colors and shapes and showy clusters of tiny, star-shaped flowers. Hundreds of cultivars, often offered with no information on species. Some need at least moderate summer water. Those listed here thrive with occasional to infrequent water in sun to part shade with good to excellent drainage. Afternoon shade in hot-summer climates. WUCOLS: L

S. adolphii, golden sedum or coppertone stonecrop, low mat 8 inches tall and 2–3 feet wide, with multiple stems bearing small rosettes of pale green, triangular leaves, tinged yellow to orange in sun, and white flowers in winter and spring. Native to the summer-rainfall, east coast of central Mexico. May be offered as *S. nussbaumerianum*, which is a similar but different plant. USDA: 9–11 Sunset: N/A

S. cauticola, cliff stonecrop, 3–6 inches tall and 12–18 inches wide, with rounded, purple-edged, blue-gray leaves on pink stems and rose-pink flowers from purplish gray buds in late summer or early fall. Dormant in winter. Native to Japan. May be offered as *Hylotelephium*. USDA: 4–9 Sunset: 1–11, 14–24

S. divergens, Pacific stonecrop, 2–5 inches tall and slowly spreading to 18 inches wide, with small, tightly bunched, round, green leaves that turn red in full sun and yellow flowers in summer. Native to rocky or gravelly soils from southeastern Alaska and western British Columbia to northern California from 4,000 to 7,500 feet. Accepts wet winters. USDA: 5–9 Sunset: N/A

S. palmeri, to 1 foot tall and spreading by trailing stems to 1–2 feet wide, with small rosettes of rounded, bluish green leaves and bright yellow flowers in winter. Native to Mexico. Performs well in shade as well as cool sun. Good choice for the Pacific Northwest and fine inland in northern California with afternoon shade. USDA: 8–10 Sunset: N/A

S. spathulifolium, Pacific stonecrop, 4–6 inches tall and 1 foot wide, with small rosettes of silvery blue-gray to gray-green leaves, sometimes tinged purple, and yellow flowers in summer. Native to rocky cliffs and bluffs from California's coastal mountains and the northern Sierra Nevada to southwestern British Columbia. 'Purpureum' has reddish purple leaves. 'Cape Blanco', a selection from southwestern Oregon, and 'Campbell Lake', from northwestern California, have chalky, gray-white leaves. USDA: 5–9 Sunset: 2–9, 14–24

^ *Sedum spathulifolium* 'Purpureum'

Sempervivum

HOUSELEEK

Succulent rosettes of leaves with pointed tips and star-shaped flowers. Each rosette flowers only once and then dies, replaced by offsets that can be removed and replanted or left to form a small colony. Native to mountains of the Mediterranean region, usually between 3,000 and 7,000 feet. Most of those offered are cultivars. Cool sun to part shade, excellent drainage, infrequent summer water. Part shade or afternoon shade in hot-summer climates. Sunset: 2–24 WUCOLS: L

S. arachnoideum, cobweb houseleek, rosette, 1–3 inches tall and wide, with small, gray-green leaves covered with fine, white hairs. Offsets to form tight colonies to 1 foot wide. Reddish purple flowers on short stems in late spring to early summer. USDA: 5–8

S. tectorum, hen and chicks, rosette, 3–4 inches tall and wide, with gray-green, sometimes purple-tipped leaves. Offsets to form mounds to 2 feet wide. Reddish purple flowers on leafy, 1-foot stems in summer. USDA: 3–8

^ *Sedum spathulifolium* 'Cape Blanco'

^ *Sempervivum tectorum*

^ *Sideritis cypria*

Sideritis cypria
SIDERITIS

Perennial, 1 foot tall and 1–2 feet wide, with upright, velvety, grayish white leaves and yellow flowers with chartreuse, cuplike bracts on upright, branching, 2- to 3-foot stems in spring and summer. Native to dry, south-facing limestone cliffs in the north coastal mountains of the island of Cyprus. Sun, excellent drainage, occasional summer water. USDA: 8–10 Sunset: N/A WUCOLS: N/A

Sisyrinchium bellum
BLUE-EYED GRASS

Perennial, 6–12 inches tall and wide, with narrow, green leaves resembling iris and purplish blue or rarely white flowers with yellow centers in spring. Summer dormant. Self-sows. Native to seasonally moist grasslands and open woodlands, inland and near the coast, usually below 6,000 feet, from

^ *Sisyrinchium bellum*

Oregon to northwestern Baja California. Sun to part shade, most well-drained soils, infrequent to no summer water. USDA: 7–10 Sunset: 4–9, 14–24 WUCOLS: L/VL

Solanum

NIGHTSHADE

Evergreen and deciduous shrubs, perennials, and vines with showy clusters of small, star-shaped flowers in summer. Most need regular water. Those listed here are evergreen and thrive with occasional to infrequent summer water in sun to part shade and well-drained soils. Drop leaves in summer if grown dry. USDA: 8b–11 WUCOLS: VL/L/M

S. umbelliferum, bluewitch, vining shrub, mounding 1–3 feet tall and wide, with densely hairy, branched, gray-green stems, small, pale gray-green leaves, and bluish purple flowers. Native to dry, rocky slopes from sea level to 5,000 feet in the foothills of southwestern Oregon south to Baja California and east to Arizona. *S. umbelliferum* var. *incanum* 'Spring Frost' has white flowers. Sunset: N/A

S. xanti, purple nightshade, vine or vining shrub, 2–4 feet tall and wide, with dark green leaves and bluish purple flowers. Native to the San Francisco Bay Area, the Sierra Nevada foothills, and coastal mountains of southern California and Baja California below 4,000 feet. 'Mountain Pride', from the foothills above Santa Barbara, has especially large, dark purple flowers. Best in part shade. Sunset: 7–9, 11, 14–24

^ *Solanum xanti* 'Mountain Pride'

^ *Sphaeralcea incana*

Sphaeralcea

GLOBE MALLOW

Perennials with softly hairy, shallowly lobed or deeply cut, green or gray-green leaves and bowl-shaped flowers. Full sun, fast-draining soils, good air circulation, occasional to infrequent or no summer water. Cut back in late winter or early spring to renew. Best in hot-summer climates. WUCOLS: L/VL

S. ambigua, desert globe mallow, 3–5 feet tall and 2–4 feet wide, with gray-green leaves and pale orange to bright red-orange flowers in spring. The subspecies *rosacea* has pink or lavender flowers. Self-sows. Native to dry, rocky slopes and sandy washes below 3,500 feet in deserts of southeastern California east to Utah and Arizona and south to Baja California. Sensitive to cold, wet winters. USDA: 6–10 Sunset: 3, 7–24

S. incana, orange globe mallow, 3–5 feet tall and 2–3 feet wide, with gray-green leaves and orange flowers in late spring into summer. Plants with deep pink flowers are available. Native to Arizona, New Mexico, Texas, and northern Mexico. More accepting of wet winters than *S. ambigua*. May spread with summer water. USDA: 7b–10 Sunset: N/A

S. munroana, Munro's globe mallow, 2–3 feet tall and wide, with pale gray-green leaves and pale pinkish orange to bright red-orange flowers in midsummer to fall. Native to the Rocky Mountains, Great Basin, and Sierra Nevada below 6,500 feet from British Columbia south through eastern Washington, Oregon, and California, and southeast to Utah and Colorado. Adaptable. USDA: 4–9 Sunset: 1–3, 7–10, 14–24

^ *Sporobolus airoides*

Sporobolus

SACATON, DROPSEED

Warm-season bunchgrasses with fine-textured foliage, good fall color, and deep roots that help to control erosion. Sun to part shade, most well-drained soils, moderate to occasional or infrequent summer water.

S. airoides, alkali sacaton, 1–2 feet tall and wide, with narrow, arching, silvery gray-green leaves that age to light tan and airy, pink-tinted midsummer flowers on 4-foot stems. Native to a wide range of habitats in much of western North America, often where there is underground or seasonal moisture. Moderate summer water and full sun. USDA: 4–9 Sunset: 1–24 WUCOLS: L/VL

S. heterolepis, prairie dropseed, 1–2 feet tall and wide, with narrow, almost threadlike, dark green leaves that turn bronzy tan in fall and airy, pinkish, late-summer flowers on 3-foot stems. Native to dry, rocky soils below 7,000 feet from southern Canada to Colorado, New Mexico, and Texas and east to the New England states. USDA: 3–8 Sunset: 1–10, 14–17 WUCOLS: N/A

S. wrightii, giant sacaton, 4–5 feet tall and 3–5 feet wide, with dark green leaves and summer-blooming flower spikes 5–6 feet tall. Native to rocky slopes and dry lake beds below 7,000 feet from eastern California to Texas and northern Mexico. 'Windbreaker', a selection from New Mexico, is 6–10 feet tall. Good hedge or screen. USDA: 5–8 Sunset: 3b, 7–16, 18–24 WUCOLS: L/VL

^ *Sporobolus wrightii*

Sporobolus wrightii 'Windbreaker' >

Stachys byzantina
LAMB'S EARS

Perennial, 6–10 inches tall and spreading by rooting stems to 4–6 feet wide, with dense rosettes of soft, velvety, silvery gray-green leaves and insignificant purplish pink flowers. Native to rocky slopes and scrublands below 6,500 feet in northeastern Turkey, Armenia, and northern Iran. Grown mostly for foliage. 'Silver Carpet' and 'Big Ears' rarely flower. Cool sun to part shade, fast drainage, occasional to infrequent summer water. USDA: 4–8 Sunset: 1–24 WUCOLS: L/M

^ *Stachys byzantina* 'Big Ears'

Stipa
FEATHER GRASS, NEEDLEGRASS

Cool-season bunchgrasses with narrow, green to gray-green leaves and airy flowers on tall, arching stems in spring. Most self-sow. Dormant or partially dormant in summer. Sun to light shade, good drainage, occasional to infrequent or no summer water. *S. tenuissima* (formerly *Nassella*) is considered a high risk for invasiveness in parts of California and Oregon.

S. arundinacea*, see *Anemanthele lessoniana

S. cernua, nodding needlegrass, 1 foot tall and 1–2 feet wide, with bluish green leaves and purplish flowers on 2-foot stems. Native to grasslands and foothill woodlands below 4,500 feet in coastal mountains of California and northern Baja California. Infrequent to no summer water. May be offered as *Nassella*. USDA: 7–10 Sunset: 7–9, 11, 14–24 WUCOLS: VL

^ *Stipa cernua*

S. gigantea, giant feather grass, 2 feet tall and wide, with gray-green leaves and purplish flowers on 6-foot stems. Native to rocky or sandy soils from Portugal and southern Spain to northern Morocco. 'Little Giant' has 3- to 4-foot flower stems. Occasional summer water. USDA: 6–10 Sunset: 4–9, 14–24 WUCOLS: N/A

^ *Stipa gigantea*

S. lepida, foothill needlegrass, 1–2 feet tall and wide, with green leaves and silvery gray flowers on 2- to 3-foot stems. Native to dry slopes below 4,000 feet in coastal mountains of California and Sierra Nevada foothills into northern Baja California. Part shade and infrequent to no summer water. May be offered as *Nassella*. USDA: 7–10 Sunset: 7–9, 11, 14–24 WUCOLS: VL

S. pulchra, purple needlegrass, 1–2 feet tall and wide, with bright green to grayish green leaves and purplish flowers on 2- to 4-foot stems. Native from coastal northern California, the Central Valley, and the western Sierra Nevada foothills south to northern Baja California. Infrequent to no summer water. May be offered as *Nassella*. USDA: 7–11 Sunset: 5–9, 11, 14–24 WUCOLS: VL

Stylomecon, see *Papaver*

Styrax redivivus

STYRAX

Deciduous shrub, multistem, slow growing to 6–10 feet tall and wide, with silvery gray bark and rounded, dark green leaves that turn yellow and orange in fall. Pendant clusters of fragrant, pure white, waxy, bell-shaped flowers in spring to early summer are followed by green fruits aging to tan. Native intermittently below 5,000 feet in southwestern Oregon, the inner Coast Ranges and Sierra Nevada foothills in northern California, and coastal mountains in southern California. Cool sun or light shade, most well-drained soils, infrequent summer water. May be offered as a subspecies of *S. officinalis*. Other *Styrax* species need regular water. USDA: 8–10 Sunset: 6–10, 14–24 WUCOLS: L/M

^ *Styrax redivivus*

^ *Symphyotrichum chilense*

Symphyotrichum chilense

PACIFIC ASTER

Perennial from rhizomes, 1–3 feet tall and 2–3 feet wide, with narrowly oval to lance-shaped, lightly hairy, green leaves and lavender or white, late-summer daisy flowers with yellow centers. Native to many habitats from British Columbia to central California, especially along the coast. Spreads rapidly by rhizomes in moist locations. 'Point Saint George', with pale lavender flowers, is 4–6 inches tall. 'Purple Haze' is 2–3 feet tall with deep purple flowers. Sun to part shade, most soils, infrequent summer water. Accepts periodic flooding. May be offered as *Aster*. USDA: 6–10 Sunset: N/A WUCOLS: L/M

Tagetes lemmonii
MEXICAN MARIGOLD

Evergreen shrub, 4–6 feet tall and 6–8 feet wide, with finely dissected, medium green, aromatic leaves and yellow-orange daisy flowers in fall and winter. Native to canyons in mountains of southern Arizona and northern Mexico between 4,000 and 8,000 feet. 'Compacta' is 2 feet tall and 3–4 feet wide. Sun to part shade, most well-drained soils, occasional summer water. USDA: 8b–11 Sunset: 8–10, 12–24 WUCOLS: L/M

^ *Tagetes lemmonii*

Tecoma stans
YELLOW BELLS

Evergreen shrub to small tree, 10–20 feet tall and 10–15 feet wide, with bright green leaves divided into many lance-shaped leaflets and showy clusters of trumpet-shaped, yellow flowers from spring to fall, followed by long, brown seedpods. Native from southern Texas to southern Arizona and south to Mexico and parts of Central and South America. Several cultivars are 6–8 feet tall, including SOLAR FLARE, with orange flowers, SUNRISE, with orange and yellow flowers, and BELLS OF FIRE, with red flowers. Sun to part shade, good drainage, moderate to occasional summer water. Needs summer heat. USDA: 9–11 Sunset: 12–13, 21–24 WUCOLS: L/M

^ *Tecoma stans* SUNRISE

Teucrium
GERMANDER

Perennials with aromatic leaves and small, pink to lavender flowers in spring and early summer. Sun to light shade, good drainage, occasional to infrequent summer water.

T. betonicum, Madeira germander, 4 feet tall and wide, with heart-shaped, sage-green leaves and tall spikes of pinkish purple flowers. Native to rocky soils and coastal bluffs on the island of Madeira off the coast of Morocco. USDA: 9–11 Sunset: N/A WUCOLS: L/M

^ *Teucrium chamaedrys*

T. chamaedrys, wall germander, 1–2 feet tall and 2–3 feet wide, with small, glossy, dark green leaves and pink to purplish flowers. Native to the Mediterranean region from Portugal to Turkey. 'Nanum' is less than 1 foot tall. May be offered as *T.* ×*lucidrys*. USDA: 5–9 Sunset: 2–24 WUCOLS: L/M

T. cossonii, Majorcan germander, 4–8 inches tall and 2–3 feet wide, with narrow, gray-green leaves, whitish beneath, and pinkish lavender flowers. Native to rocky soils on the Mediterranean island of Mallorca, Spain. Needs perfect drainage. May be offered as *T. majoricum*. USDA: 8–11 Sunset: 7–9, 14–24 WUCOLS: L/VL

T. fruticans, bush germander, 4–8 feet tall and wide, with small, gray-green leaves on downy, white stems and tiny lavender-blue flowers. Native to rocky soils in the western and central Mediterranean region from Portugal to Italy. 'Azureum' is 3–4 feet tall and wide with darker blue flowers. USDA: 8–10 Sunset: 4–24 WUCOLS: L/M

T. ×lucidrys, see T. chamaedrys

^ *Teucrium fruticans*

^ *Teucrium fruticans 'Azureum'*

^ *Thalictrum fendleri*

^ *Thymus citriodorus*

Thalictrum fendleri var. *polycarpum*

MEADOW RUE

Perennial, 3–5 feet tall, with finely cut, bluish green leaves and small, pendant, pink or greenish yellow flowers on tall, sometimes purplish stems. Summer dormant. Native to seasonally moist woodland and forest openings from sea level to 10,000 feet from Oregon and Wyoming south to California, Texas, and northern Mexico. Part to full shade, humusy soils, occasional to infrequent or no summer water. Accepts periodic flooding. May be offered as *T. polycarpum*. USDA: 4–10 Sunset: N/A WUCOLS: M

Thymus

THYME

Perennials, prostrate to mounding with green to gray-green, aromatic leaves and masses of tiny flowers in spring or summer. Many named cultivars with varying leaf and flower colors. Full sun to part shade, excellent drainage, moderate to occasional summer water. Shear taller plants occasionally for best form and more flowers. Sunset: 1–24 WUCOLS: L/M

T. citriodorus, lemon thyme, 4 inches tall and 1–2 feet wide, with medium green leaves and pale lilac flowers. Native to the western Mediterranean region. Long considered a hybrid of *T. vulgaris* and *T. pulegioides* known as *T.* ×*citriodorus* and still widely offered as such. USDA: 5–11

^ Thymus serpyllum

T. pseudolanuginosus, woolly thyme, flat mat 1–2 inches tall and spreading 1 foot wide, with gray-green leaves on hairy stems and pink flowers. Native to the western Mediterranean region. Good between pavers or draped over walls. May be offered as *T. praecox* subsp. *britannicus*. USDA: 6–8

T. serpyllum, creeping thyme or mother-of-thyme, 2–3 inches tall and spreading 1–3 feet wide, with dark bluish green leaves and deep pink to purple flowers. Native to open, sandy or rocky soils in parts of northern and central Europe. Needs moderate summer water. May be offered as *T. praecox* subsp. *britannicus* or subsp. *arcticus*. USDA: 4–8

Trachycarpus

WINDMILL PALM

Palms with fan-shaped fronds on long stalks. Sun to part shade, good drainage, moderate to occasional summer water. Good choice for the Pacific Northwest. USDA: 8–10 Sunset: 4–24 WUCOLS: L/M

^ Trachycarpus fortunei

T. fortunei, windmill palm, 20–30 feet tall, with dark green fronds on 2- to 3-foot stalks, drooping clusters of small, greenish yellow flowers, and dark blue fruit on female plants. Leaves drop and rough, fibrous leaf bases persist on the trunk. Native to mountains of central and southern China at elevations up to 7,800 feet. 'Wagnerianus', dwarf Chusan palm, is 12–20 feet tall, with smaller leaves on shorter stalks. May be offered as *T. wagnerianus*.

T. takil, Kumaon palm, 30–40 feet tall, with larger fronds consisting of more leaflets than *T. fortunei*. Native to the Himalayan foothills between 7,000 and 9,000 feet. One of the most cold-hardy fan palms.

^ *Trichostema lanatum*

Trichostema lanatum

WOOLLY BLUE CURLS

Evergreen shrub, 3–4 feet tall and 4–5 feet wide, with narrow, glossy, green leaves, woolly white beneath, and foot-long spikes of blue flowers from woolly, purplish buds in late spring through summer. Native to dry slopes in coastal mountains of southern California and northern Baja California. Short lived and not always easy in cultivation. Sun to part shade, perfect drainage, no summer water. 'Midnight Magic', a hybrid of *T. lanatum* and *T. purpusii*, a species from Mexico, may be more adaptable to garden conditions; it needs moderate to occasional water. USDA: 9–11 Sunset: 14–24 WUCOLS: L/VL

^ *Triteleia ixioides*

Triteleia

TRITELEIA

Perennials from corms, with one to a few grasslike leaves and a loose, open cluster of upfacing flowers on tall, leafless stems in late spring or early summer. Most are native to sandy or gravelly soils in sunny forest or woodland openings in central and northern California and southwestern Oregon. Spread slowly by offsets. Summer dormant. Sun to part shade, most well-drained soils, no summer water. Need ample spring moisture to bloom well. May be offered as *Brodiaea*.

T. ixioides, golden brodiaea, with yellow flowers. USDA: 6–9 Sunset: 3–9, 14–24 WUCOLS: N/A

T. laxa, Ithuriel's spear, with purplish blue flowers. Commonly available as 'Queen Fabiola'. USDA: 7–10 Sunset: 5–9, 14–24 WUCOLS: VL

Vauquelinia

ROSEWOOD

Evergreen shrubs or small trees, usually multitrunk, with leathery, lance-shaped, green leaves with finely toothed or serrated margins, flat-topped clusters of creamy white flowers in spring, and persistent woody seed capsules. Sun to part shade, most well-drained soils, infrequent summer water. Best inland. Need summer heat. Sunset: 10–13

V. californica, Arizona rosewood, 8–20 feet tall and 6–15 feet wide. Native to dry slopes and canyons from 2,500 to 5,000 feet in southern Arizona, New Mexico, and northwestern Mexico. USDA: 8–10 WUCOLS: L/VL

V. corymbosa, slimleaf rosewood, 10–15 feet tall and 8–12 feet wide, with longer, more narrow leaves than *V. californica*. The subspecies *angustifolia*, Chisos rosewood, has extremely narrow leaves that are more upright than the species. Native to southwestern Texas and northeastern Mexico, usually between 3,500 to 6,500 feet. USDA: 7b–9 WUCOLS: L

^ *Vauquelinia corymbosa* subsp. *angustifolia*

Verbena

VERBENA

Perennials with showy clusters of small, tubular flowers in spring through early fall. Most plants offered are hybrids that need regular water. Those listed here do well with occasional to infrequent summer water in sun to part shade with good drainage and good air circulation. USDA: 7–10

V. bonariensis, purple top, 2–4 feet tall and 1–3 feet wide, with narrow, toothed, dark green leaves, sparsely displayed on upright stems, and small, rounded clusters of purple flowers. Native to Brazil and Argentina. 'Lollipop' is smaller. Self-sows, sometimes aggressively. High risk of invasiveness in California, especially in moist soils. Sunset: 8–24 WUCOLS: L/VL/M

V. lilacina*, see *Glandularia

^ *Verbena bonariensis*

V. rigida, sandpaper verbena, 1–2 feet tall and 4–5 feet wide, with rough-textured, dark green leaves with widely spaced, sharply pointed, marginal teeth and tight clusters of lavender flowers on tall stems. Native to Brazil, Bolivia, and Argentina. Spreads widely, especially in moist soils. Cut back hard after flowering to renew. May be offered as a variety of *V. bonariensis* or as *V. venosa*. Sunset: 3–24 WUCOLS: L/M

Vitex
CHASTE TREE

Deciduous and evergreen shrubs, fast growing, with grayish green, aromatic leaves, usually divided into oval to lance-shaped leaflets, and spikes of small, lavender-blue or purplish flowers. Sun, most soils, moderate to occasional summer water. Self-sow and can be weedy or invasive.

V. agnus-castus, chaste tree, deciduous, 10–15 feet tall and 15–20 feet wide, with showy, mostly upright spikes of small, fragrant flowers from summer to fall. Native to streambanks and other damp places in southern Europe and western Asia. BLUE DIDDLEY ('SMVACBD'), a selection from Michigan, is half the size of the species. USDA: 7b–11 Sunset: 4–24 WUCOLS: L

V. trifolia, Arabian lilac or blue vitex, evergreen, 10–15 feet tall and wide, with short spikes of flowers in summer through fall and into winter. Native to coastal areas from eastern Africa east through Asia and south to Australia. Usually available as 'Purpurea'. May be offered as variety *purpurea* or as *V.* 'Fascination'. Best near the coast. USDA: 9–11 Sunset: N/A WUCOLS: L/M

< *Vitex agnus-castus*

^ *Vitex trifolia* 'Purpurea'

Vitis
GRAPE

Deciduous, fast-growing vines, climbing by tendrils that attach themselves tightly to anything they encounter, including nearby shrubs and trees. Prune at any time to control spread. Leaves are large, soft, and slightly lobed and serrated, often with brilliant fall colors. Berries are smaller than those grown for edible fruit but attractive to many birds. Sun to part shade, most soils, occasional to infrequent or no summer water. USDA: 7–10

V. californica, California wild grape, to 30 feet or more, with heart-shaped to round, green leaves that turn bright red and yellow in fall. Native to seasonally damp places in coastal mountains of southwestern Oregon and northern California, the Sacramento Valley, and the foothills of the Transverse Ranges in southern California. 'Walker Ridge', a selection found growing on serpentine soils in northern California, is smaller, to 10 feet. Sunset: 4–24 WUCOLS: L/VL/M

V. girdiana, desert grape, similar to *V. californica* but with stems, leaf undersides, and new leaves covered in white hairs, giving the plant a silvery gray appearance. Native to streambanks and canyons from sea level to 4,100 feet in southern California and Baja California. Good choice for inland locations. Sunset: N/A WUCOLS: L/M

V. 'Roger's Red', a hybrid between *V. californica* and the wine grape cultivar *V. vinifera* 'Alicante Bouschet', has especially brilliant red-orange fall coloring. Sunset: 4-24 WUCOLS: L/M

Westringia fruticosa

COAST ROSEMARY

Evergreen shrub, 4-6 feet tall and 5-10 feet wide, with dark gray-green, narrowly linear leaves and small, white flowers almost year-round in mild climates. Native to sandy soils along the coast in New South Wales, Australia. BLUE GEM ('WES03') has olive-green leaves and blue-purple flowers. 'Smokey' and 'Morning Light' have variegated leaves and white flowers. GREY BOX ('WES04'), with gray-green leaves and white flowers, is 3 feet tall and wide. Sun to light shade, good drainage, occasional to infrequent summer water. May be offered as *W. rosmariniformis*. Good seaside plants. USDA: 9-10 Sunset: 8-9, 14-24 WUCOLS: L

W. 'Wynyabbie Gem', 6 feet tall and wide, with gray-green leaves and lavender flowers, is a hybrid between *W. fruticosa* and *W. eremicola*, an inland species from southeastern Australia. 'Wynyabbie Highlight', a sport of 'Wynyabbie Gem', has variegated leaves.

^ *Vitis* 'Roger's Red'

^ *Westringia fruticosa* 'Smokey'

^ *Westringia fruticosa* 'Morning Light'

^ *Wyethia angustifolia*

^ *Xanthorrhoea quadrangulata*

Wyethia angustifolia

MULE'S EARS

Perennial, 1 foot tall and 1–2 feet wide, low rosette of long, lance-shaped, green leaves and large, bright yellow sunflowers on leafless, 2-foot stems in spring. Dies back to the ground in winter. Native to grassy slopes and forest openings, often along streams, from southwestern Washington to central California and from the coast to the northern Sierra Nevada foothills. Full sun, most soils, occasional to infrequent summer water. Not at its best in southern California. USDA: 5–8 Sunset: N/A WUCOLS: L

Xanthorrhoea

GRASS TREE

Perennials with stiff, narrow, grasslike leaves radiating out from a slowly forming stem or trunk. At maturity, but not every year, thousands of tiny, white, star-shaped flowers on a tall, thin stalk in spring. Slow growing and long lived. Not often available but worth the search. Sun to light shade, well-drained soils, infrequent to no summer water. USDA: 9–11 Sunset: 12–24 WUCOLS: L

X. preissii, 2- to 4-foot rosette of blue-green leaves atop a 4- to 8-foot stem and flowers on a 5- to 10-foot stalk. Native to rocky soils, often near streams, on coastal plains and in adjacent forests of southwestern Western Australia.

X. quadrangulata, 3- to 4-foot rosette of blue-green leaves on a branching stem to 6 feet tall and flowers on a 6- to 12-foot stalk. Native to rocky outcrops in South Australia.

Yucca
YUCCA

Succulent rosettes of strap-shaped or sword-shaped leaves, some with stems or trunks and some without, and creamy white flowers on branched or unbranched stalks in spring or early summer. Most yuccas do not die after flowering, but rosettes may not flower every year. Some yuccas prefer moderate to regular water. Those listed here thrive with infrequent to no summer water with fast drainage and in sun.

Y. baccata, banana yucca or datil, stemless or short-stemmed rosette to 3 feet tall and 5 feet wide, with rigid, spine-tipped, bluish green leaves and purple-tinged, creamy white flowers on a 3- to 6-foot stalk. Offsets to form large clusters. Native to arid and semi-arid grasslands and woodlands from southeastern California north to Utah and south to western Texas and northern Mexico. USDA: 6–11 Sunset: 1–3, 7, 9–14, 18–24 WUCOLS: VL

Y. glauca, soapweed yucca, stemless or short-stemmed rosette to 3–4 feet tall and wide, with stiff, narrow, grayish green leaves and pendant, greenish white flowers on a 3- to 4-foot stalk. Native to dry, rocky soils in the Great Plains from southern Alberta and Saskatchewan south to New Mexico, Texas, and northeastern Mexico. USDA: 5–11 Sunset: 1–13 WUCOLS: L/VL

Y. linearifolia, linear-leaved yucca, spherical rosette to 2–3 feet tall and wide, with narrow, green leaves, slowly forming a 4- to 6-foot stem. White flowers on a 3-foot stalk. Native to the Chihuahuan Desert of northeastern Mexico and southwestern Texas. May be offered as *Y. rostrata* var. *linearis*. USDA: 7b–10 Sunset: N/A WUCOLS: VL

Y. rostrata, Big Bend yucca, spherical rosette to 4–6 feet tall and wide, with stiff, narrow, sharp-pointed, silvery gray-blue leaves, slowly forming a thick stem 10–12 feet tall. Creamy white flowers

^ *Yucca rostrata*

^ *Yucca rostrata* 'Sapphire Skies'

on a 4-foot stalk. Native to rocky slopes and ridges in southwestern Texas and northeastern Mexico. 'Sapphire Skies' is a powder-blue selection made in Oregon from seed collected in Mexico. USDA: 7b–11 Sunset: 7–24 WUCOLS: VL

Y. schidigera, Mojave yucca, rosette to 3–4 feet tall and wide, with thick, rigid, yellow-green to blue-green leaves with marginal fibers and a sharp terminal spine. Slowly forms a stem or several stems to 5–15 feet tall. Creamy white flowers on a 1- to 4-foot stalk. Native to dry, rocky slopes and flats from sea level to 8,000 feet in southern California and northern Baja California east to New Mexico. USDA: 9–10 Sunset: 7–16, 18–24 WUCOLS: VL

Y. whipplei*, see *Hesperoyucca

Yucca schidigera >

Plants for Special Places

^ *Ceanothus* 'Ray Hartman', at center, can be pruned up as a small tree. Coffeeberry, at right, can be pruned up and underplanted with wildflowers.

SMALL SPACES, hot sites, dry shade, windy coastal locations, and other site conditions present challenges for the gardener or garden designer. Whatever the need, there is usually a plant to fill it.

TREES FOR SMALL GARDENS

Many gardens today do not have space for even one large tree. Listed here are some small to medium-sized trees and large shrubs that can be trained as single- or multitrunk trees.

Trees

Acacia boormanii, *A. cognata*, *A. covenyi*, acacia

Aesculus californica, California buckeye

Agonis flexuosa 'Jervis Bay Afterdark', peppermint willow

Brahea edulis, Guadalupe palm

Cercis canadensis var. *texensis*, *C. occidentalis*, redbud

Chilopsis linearis, desert willow

×*Chitalpa tashkentensis*, chitalpa

Eucalyptus archeri, *E. forrestiana*, *E. perriniana*, eucalyptus

Hesperocyparis forbesii, tecate cypress

Juniperus scopulorum, Rocky Mountain juniper

Lagerstroemia cultivars, crape myrtle

Lyonothamnus floribundus subsp. *aspleniifolius*, Santa Cruz Island ironwood

Melaleuca linariifolia, *M. nesophila*, melaleuca

Olea europaea, *O. europaea* 'Wilsonii' olive

Parkinsonia 'Desert Museum', palo verde

Parrotia persica, Persian ironwood

Pinus contorta, *P. edulis*, *P. monophylla*, *P. mugo*, pine

Trachycarpus fortunei, windmill palm

Shrubs

Acca sellowiana, pineapple guava

Arbutus unedo, strawberry tree

Arctostaphylos glauca, *A. manzanita*, *A.* 'Monica', manzanita

Banksia speciosa, showy banksia

Callistemon 'Cane's Hybrid', *C. citrinus*, *C. viminalis*, bottlebrush

Ceanothus arboreus, *C.* 'Concha', *C.* 'Ray Hartman', *C. thyrsiflorus* var. *thyrsiflorus* 'Snow Flurry', wild lilac

Cercocarpus betuloides, *C. ledifolius*, mountain mahogany

Comarostaphylis diversifolia, summer holly

Cotinus coggygria, *C.* 'Grace', *C. obovatus*, smoke tree

Dodonaea viscosa, hopbush

Hakea francisiana, *H. laurina*, *H. petiolaris*, hakea

Heteromeles arbutifolia, toyon

Laurus nobilis, bay laurel

Leptospermum 'Dark Shadows', *L. laevigatum*, *L. lanigerum*, tea tree

Malosma laurina, laurel sumac

Morella californica, Pacific wax myrtle

^ *Arctostaphylos* 'Dr. Hurd' can be pruned up as an attractive small tree.

Oemleria cerasiformis, osoberry

Osmanthus ×fortunei, *O. fragrans*, *O. heterophyllus*, tea olive

Pistacia lentiscus, mastic

Prunus ilicifolia, hollyleaf cherry

Rhamnus alaternus, Italian buckthorn

Rhaphiolepis MAJESTIC BEAUTY ('Montic'), India hawthorn

Sambucus cerulea, elderberry

Vauquelinia californica, *V. corymbosa*, rosewood

Vitex agnus-castus, *V. trifolia*, chaste tree

HEDGES AND SCREENS

Some trees and shrubs are especially well suited to the task of screening for privacy or obscuring undesired views and also may serve to buffer a site from winds.

Acacia boormanii, *A. covenyi*, acacia

Acca sellowiana, pineapple guava

Adenanthos sericeus, woollybush

Arbutus unedo, strawberry tree

Arctostaphylos 'Austin Griffiths', *A. bakeri* 'Louis Edmunds', *A.* 'Dr. Hurd', *A. glauca*, *A.* 'Lester Rowntree', *A. manzanita*, *A.* 'Monica', manzanita

Baeckea virgata, baeckea

Banksia ericifolia, *B. speciosa*, banksia

Buxus sempervirens, boxwood

Callistemon 'Cane's Hybrid', *C. citrinus*, *C. viminalis*, bottlebrush

Calothamnus quadrifidus, one-sided bottlebrush

Ceanothus arboreus, *C.* 'Concha', *C.* 'Cynthia Postan', *C.* 'Julia Phelps', *C. thyrsiflorus* var. *thyrsiflorus* 'Snow Flurry', *C.* 'Victoria', wild lilac

Comarostaphylis diversifolia, summer holly

Dodonaea viscosa, hopbush

Elaeagnus pungens, silverberry

Frangula californica, coffeeberry

Garrya elliptica, *G. fremontii*, *G. ×issaquahensis*, silktassel

Grevillea 'Kings Fire', *G.* 'Neil Bell', grevillea

Hesperocyparis forbesii, cypress

Heteromeles arbutifolia, toyon

Laurus nobilis, bay laurel

Leptospermum 'Dark Shadows', *L. lanigerum*, *L. namadgiensis*, *L. scoparium* 'Washington Park', tea tree

Malosma laurina, laurel sumac

Melaleuca elliptica, *M. incana*, melaleuca

Morella californica, wax myrtle

Myrtus communis, myrtle

Olea europaea 'Skylark Dwarf', dwarf olive

Olearia ×haastii, daisy bush

Osmanthus species and cultivars, tea olive

Pistacia lentiscus, mastic

Prunus ilicifolia, hollyleaf cherry

Rhamnus alaternus, *R. ilicifolia*, buckthorn

Rhus integrifolia, *R. ovata*, sumac

Vauquelinia californica, rosewood

PLANTS FOR HOT SITES

Many plants cannot take full sun in hot-summer climates or require considerable water if grown in a hot location. Plants listed here prefer or accept hot sun with moderate to occasional or infrequent water once established.

Trees

Acacia boormanii, *A. covenyi*, acacia

Brahea armata, blue hesper palm

Chilopsis linearis, desert willow

×*Chitalpa tashkentensis*, chitalpa

Hesperocyparis arizonica, cypress

Lagerstroemia cultivars, crape myrtle

Olea europaea, olive

Parkinsonia 'Desert Museum', palo verde

Pistacia chinensis, *P. lentiscus*, pistache, mastic

Quercus douglasii, *Q. wislizeni*, oak

Shrubs

Acacia redolens, prostrate acacia

Artemisia tridentata, big sagebrush

Atriplex canescens, *A. lentiformis*, saltbush

Baccharis 'Centennial', *B.* 'Starn', coyote brush

Calliandra californica, *C. eriophylla*, fairy duster

Calothamnus quadrifidus, one-sided bottlebrush

Cercocarpus ledifolius, mountain mahogany

Chamelaucium uncinatum, Geraldton waxflower

Cistus cultivars, rockrose

Condea emoryi, desert lavender

Dalea frutescens, *D. greggii*, dalea

Dasylirion longissimum, *D. texanum*, *D. wheeleri*, desert spoon

Dodonaea viscosa, hopbush

Elaeagnus pungens, *E.* ×*submacrophylla*, silverberry

Encelia farinosa, brittlebush

Eriogonum fasciculatum, buckwheat

Fallugia paradoxa, Apache plume

Forestiera pubescens, desert olive

Justicia californica, *J. spicigera*, justicia

Leucophyllum frutescens, *L. langmaniae*, Texas ranger

Malacothamnus fremontii, bush mallow

Nolina species, bear grass

Peritoma arborea, bladderpod

Rosmarinus officinalis, rosemary

Salvia clevelandii, *S. leucophylla*, sage

Vauquelinia californica, *V. corymbosa*, rosewood

Vitex agnus-castus, *V. trifolia*, chaste tree

Perennials, Annuals, Grasses

Achillea millefolium, yarrow

Allium cristophii, wild onion

Berlandiera lyrata, chocolate flower

Bidens ferulifolia, bur marigold

Bouteloua gracilis, blue grama grass

Dicliptera sericea, firecracker plant

Eriogonum umbellatum, sulfur buckwheat

Glandularia gooddingii, Mojave verbena

Melampodium leucanthum, blackfoot daisy

^ Palo verde (*Parkinsonia* 'Desert Museum') with agaves and golden barrel cactus in southern California

Perennials, Annuals, Grasses, *continued*

Mentzelia lindleyi, blazing star

Muhlenbergia capillaris, *M. dubia*, *M. lindheimeri*, *M. reverchonii*, muhly

Nolina species, bear grass

Perovskia atriplicifolia, Russian sage

Romneya coulteri, Matilija poppy

Sphaeralcea ambigua, *S. incana*, globe mallow

Succulents

Agave havardiana, *A. parryi*, agave

Aloe arborescens, *A.* 'Blue Elf', aloe

Delosperma cooperi, *D. floribundum*, ice plant

Furcraea macdougallii, Macdougall's century plant

Hesperaloe parviflora, red yucca

Hesperoyucca whipplei, chaparral yucca

Yucca species, yucca

Vines

Bougainvillea cultivars, bougainvillea

PLANTS FOR DRYISH SHADE

Many plants appreciate light shade and most of these will do well with less water when grown in part shade. Plants listed here are content in light shade with infrequent to no water once established.

Shrubs

Arbutus unedo, *A. unedo* 'Compacta', strawberry tree

Arctostaphylos 'Pacific Mist', manzanita

Berberis aquifolium var. *repens*, *B. nevinii*, mahonia

Buxus sempervirens, boxwood

Carpenteria californica, bush anemone

Comarostaphylis diversifolia, summer holly

Correa species and cultivars, Australian fuchsia

Elaeagnus pungens, silverberry

Frangula californica, coffeeberry

Garrya elliptica, *G. fremontii*, silktassel

Holodiscus discolor, oceanspray

Mahonia eurybracteata, threadleaf mahonia

Morella californica, Pacific wax myrtle

Philadelphus lewisii, mock orange

Rhamnus alaternus, *R. crocea*, buckthorn

Rhus integrifolia, *R. ovata*, sumac

Ribes speciosum, *R. viburnifolium*, gooseberry, currant

Rubus rolfei, Taiwan bramble

Perennials, Annuals, Grasses

Acanthus mollis, *A. spinosus*, bear's breech

Anemanthele lessoniana, New Zealand wind grass

Bergenia crassifolia, bergenia

Beschorneria yuccoides, amole

Calamagrostis foliosa, Mendocino reedgrass

Carex praegracilis, *C. tumulicola*, sedge

Clarkia species, clarkia

Cyclamen coum, *C. hederifolium*, cyclamen

Diplacus aurantiacus, sticky monkeyflower

Dryopteris arguta, coastal wood fern

Elymus condensatus, giant wild rye

Epimedium species, epimedium

Euphorbia characias, spurge

Festuca californica, California fescue

Fragaria vesca, woodland strawberry

Geranium macrorrhizum, bigroot geranium

Heuchera species and cultivars, coral bells

Iris douglasiana, *I. tenax*, iris

Juncus patens, California gray rush

^ *Heuchera* 'Rosada' with agaves and iris thrive in dry shade.

Perennials, Annuals, Grasses, *continued*

Keckiella cordifolia, bush penstemon

Lepechinia fragrans, *L. hastata*, pitcher sage

Lomandra longifolia BREEZE ('LM300'), mat rush

Lonicera hispidula, *L. subspicata*, honeysuckle

Melica californica, *M. imperfecta*, *M. torreyana*, melic grass

Muhlenbergia rigens, deer grass

Phacelia bolanderi, woodland phacelia

Polypodium californicum, California polypody

Polystichum munitum, western sword fern

Salvia spathacea, hummingbird sage

Stipa pulchra, purple needlegrass

Thalictrum fendleri var. *polycarpum*, meadow rue

Succulents

Agave attenuata, *A. bracteosa*, agave

Crassula multicava, *C. ovata*, crassula

Echeveria cultivars, hen and chicks

Graptopetalum paraguayense, ghost plant

Sedum palmeri, *S. spathulifolium*, stonecrop

Sempervivum arachnoideum, houseleek

Vines

Aristolochia californica, California pipevine

Hardenbergia comptoniana, lilac vine

PLANTS FOR COASTAL GARDENS

Wind and salt-laden air and soil can stress plants not adapted to seaside environments. Plants listed here do well along the coast, although not necessarily in beach sand or directly exposed to salt spray.

Trees

Acacia boormanii, *A. cognata*, *A. covenyi*, acacia

Agonis flexuosa, peppermint willow

Brahea edulis, Guadalupe palm

Chamaerops humilis, Mediterranean fan palm

Corymbia citriodora, *C. ficifolia*, gum tree

Cupressus sempervirens, cypress

Eucalyptus forrestiana, eucalyptus

Melaleuca linariifolia, *M. nesophila*, melaleuca

Pinus contorta, *P. torreyana*, pine

Rhaphiolepis MAJESTIC BEAUTY ('Montic'), India hawthorn

Trachycarpus fortunei, windmill palm

Shrubs

Acacia redolens, prostrate acacia

Adenanthos cuneatus, *A. sericeus*, woollybush

Aloysia citrodora, lemon verbena

Arbutus unedo, strawberry tree

Arctostaphylos edmundsii, *A. hookeri*, *A. uva-ursi*, manzanita

Artemisia pycnocephala, sandhill sage

Baccharis magellanica, *B. pilularis* 'Pigeon Point', *B. pilularis* 'Twin Peaks 2', coyote brush

Banksia speciosa, banksia

Brachyglottis greyi, daisy bush

Bupleurum fruticosum, shrubby hare's ear

Callistemon species, bottlebrush

Calothamnus quadrifidus, one-sided bottlebrush

Ceanothus gloriosus var. *gloriosus*, *C. maritimus*, *C. thyrsiflorus* var. *griseus*, *C. thyrsiflorus* var. *thyrsiflorus* 'Snow Flurry', wild lilac

Cistus species and cultivars, rockrose

Constancea nevinii, Nevin's woolly sunflower

Convolvulus cneorum, bush morning glory

Correa species and cultivars, Australian fuchsia

Dendromecon harfordii, island tree poppy

Dodonaea viscosa, hopbush

Dorycnium hirsutum, canary clover

Elaeagnus pungens, silverberry

Encelia californica, brittlebush

Erica canaliculata, Christmas heather

Erigeron glaucus, seaside daisy

Eriogonum arborescens, *E. giganteum*, *E. latifolium*, buckwheat

Euryops pectinatus, euryops

Gambelia speciosa, island snapdragon

Garrya elliptica, silktassel

Grevillea lanigera, grevillea

Hakea suaveolens, hakea

Heteromeles arbutifolia, toyon

Juniperus conferta, shore juniper

Laurus nobilis, bay laurel

Lepechinia fragrans, island pitcher sage

Shrubs, *continued*

Leptospermum scoparium, New Zealand tea tree

Leucophyta brownii, cushion bush

Lonicera nitida, boxleaf honeysuckle

Lyonothamnus floribundus subsp. *aspleniifolius*, Santa Cruz Island ironwood

Malosma laurina, laurel sumac

Malva assurgentiflora, island mallow

Melaleuca elliptica, *M. incana*, melaleuca

Morella californica, Pacific wax myrtle

Myrsine africana, African boxwood

Olearia ×haastii, daisy bush

Peritoma arborea, bladderpod

Perityle incana, Guadalupe Island rock daisy

Phlomis fruticosa, Jerusalem sage

Prunus ilicifolia subsp. *lyonii*, Catalina cherry

Rhagodia spinescens, Australian saltbush

Rhamnus alaternus, Italian buckthorn

Rhaphiolepis indica, *R. umbellata*, India hawthorn

Rhus integrifolia, lemonade berry

Ribes viburnifolium, Catalina currant

Rosa banksiae, *R. minutifolia*, rose

Rosmarinus officinalis, rosemary

Salvia apiana, *S. greggii*, *S. leucophylla*, sage

Westringia fruticosa, coast rosemary

Perennials

Achillea filipendulina, *A. millefolium*, *A. millefolium* 'Island Pink', yarrow

Armeria maritima, sea thrift

Asteriscus maritimus, gold coin

Ballota pseudodictamnus, Grecian horehound

Dymondia margaretae, silver carpet

Erigeron glaucus, *E. glaucus* 'Cape Sebastian', *E.* 'W.R.', seaside daisy

Eriogonum arborescens, *E. fasciculatum*, *E. giganteum*, buckwheat

Eriophyllum staechadifolium, seaside woolly sunflower

Eryngium alpinum, *E. amethystinum*, *E. planum*, sea holly

Erysimum concinnum, *E. franciscanum*, wallflower

Eschscholzia californica, California poppy

Fragaria chiloensis, coastal strawberry

Glaucium flavum, yellow horned poppy

Helianthemum nummularium, sunrose

Iris douglasiana, Douglas iris

Leptosyne gigantea, giant coreopsis

Lessingia filaginifolia, California aster

Nepeta ×faassenii, catmint

Phacelia californica, rock phacelia

Phormium species and cultivars, flax

Romneya coulteri, Matilija poppy

Sideritis cypria, sideritis

Symphyotrichum chilense, Pacific aster

Teucrium fruticans, bush germander

Grasses and Grasslike Plants

Calamagrostis foliosa, Mendocino reedgrass

Carex flacca, *C. praegracilis*, *C. tumulicola*, sedge

Elegia elephantina, *E. tectorum*, Cape rush

Elymus condensatus 'Canyon Prince', giant wild rye

Festuca glauca, blue fescue

Helictotrichon sempervirens, blue oat grass

Juncus patens, California gray rush

Muhlenbergia rigens, deer grass

Stipa pulchra, purple needlegrass

^ Coyote brush (*Baccharis pilularis* 'Pigeon Point') and brittlebush (*Encelia californica*) grow on a hillside in a coastal California garden.

Succulents

Agave species, agave

Aloe species, aloe

Crassula ovata, jade plant

Delosperma cooperi, *D. floribundum*, ice plant

Dudleya brittonii, *D. caespitosa*, *D. farinosa*, dudleya

Echeveria species and cultivars, hen and chicks

Furcraea foetida, *F. macdougallii*, furcraea

Sedum species and cultivars, stonecrop

Sempervivum tectorum, houseleek

What Plant Where:
A Guide to
Plant
Selection

PLANTS ARE ARRANGED HERE by category (trees, shrubs, and so on) and by characteristics such as evergreen or deciduous, sun or shade, and season of bloom or of active growth.

The distinction between evergreen and deciduous is not as clear-cut in summer-dry climates as it is in climates with summer rain. Some plants that do not drop their leaves in winter will drop some or all leaves in extreme summer drought. Some but not all of these will retain their leaves if provided with summer water. Other plants that are deciduous and dormant in colder winters will be evergreen in mild-winter climates, sometimes even blooming almost year-round.

Grasses are described as cool-season or warm-season. Cool-season grasses start fresh growth with fall rains, flower in early spring to early summer, and then set seed and go partially or fully dormant. Warm-season grasses are mostly dormant through the winter, develop most of their growth in summer, and flower in mid- to late summer.

Cool-season grasses tend to be from coastal plant communities where both winters and summers are mild. Warm-season grasses tend to be from inland areas, the mountains, or deserts. Most warm-season grasses do best with some heat during the growing season, and many tolerate considerable winter cold. In the garden, occasional summer watering will extend or reactivate the growing season of some cool-season grasses, but warm-season grasses seldom grow actively in winter, regardless of climate or watering regime.

< Strawberry trees (*Arbutus unedo*), phormiums, and native grasses provide a lush composition and require little or no summer water.

TREES

			Exposure		
Genus	**Deciduous**	**Evergreen**	**Sun**	**Part Shade**	**Shade**
Acacia		x	x	x	
Aesculus	x		x	x	
Agonis		x	x	x	
Arbutus		x	x	x	
Brahea		x	x	x	
Calocedrus		x	x	x	
Cercis	x		x	x	
Chamaerops		x	x		
Chilopsis	x		x		
Chitalpa	x		x		
Corymbia		x	x		
Eucalyptus		x	x		
Hesperocyparis		x	x		
Jubaea		x	x	x	
Juniperus		x	x	x	
Lagerstroemia	x		x		
Laurus		x	x	x	
Lyonothamnus		x	x	x	
Melaleuca		x	x		
Olea		x	x		
Parkinsonia	x		x		
Parrotia	x		x	x	
Pinus		x	x	x	
Pistacia	x	x	x		
Quercus	x	x	x	x	
Trachycarpus		x	x	x	

SHRUBS

			Exposure		
Genus	**Deciduous**	**Evergreen**	**Sun**	**Part Shade**	**Shade**
Abutilon		x	x	x	
Acacia		x	x	x	
Acca		x	x	x	
Adenanthos		x	x	x	
Aloysia		x	x		
Alyogyne		x	x	x	
Arctostaphylos		x	x	x	
Artemisia		x	x		
Atriplex		x	x		
Baccharis		x	x		
Baeckea		x	x		
Banksia		x	x	x	
Berberis	x	x		x	x
Brachyglottis		x	x		
Bupleurum		x	x	x	
Buxus		x		x	x
Calliandra		x	x	x	
Callistemon		x	x		
Calluna		x	x	x	
Calothamnus		x	x		
Camellia		x		x	
Carpenteria		x	x	x	x
Ceanothus		x	x	x	
Cercis	x			x	
Cercocarpus	x	x	x	x	
Chaenomeles	x		x	x	
Chamelaucium		x	x		
Choisya		x		x	
Cistus		x	x		
Comarostaphylis		x		x	
Condea		x	x		

			Exposure		
Genus	**Deciduous**	**Evergreen**	**Sun**	**Part Shade**	**Shade**
Constancea		x	x	x	
Convolvulus		x	x		
Correa		x	x	x	x
Cotinus	x		x		
Dalea		x	x		
Dendromecon		x	x	x	
Dodonaea		x	x	x	
Dorycnium		x	x		
Elaeagnus		x	x	x	x
Encelia		x	x		
Erica		x	x	x	
Euryops		x	x		
Fallugia		x	x		
Forestiera	x		x	x	
Frangula		x	x	x	x
Fremontodendron		x	x		
Gambelia		x	x	x	
Garrya		x	x	x	x
Grevillea		x	x	x	
Hakea		x	x		
Halimium		x	x		
Hebe		x	x	x	
Helianthemum		x	x		
Heteromeles		x	x	x	
Holodiscus	x			x	x
Jasminum	x	x	x	x	
Juniperus		x	x	x	
Justicia	x	x	x	x	
Keckiella		x	x	x	x
Leonotis		x	x		
Lepechinia		x	x	x	
Leptospermum		x	x		
Leucadendron		x	x		
Leucophyllum		x	x		
Leucophyta		x	x		
Lonicera	x	x	x	x	
Mahonia		x		x	x

Genus	Deciduous	Evergreen	Exposure		
			Sun	Part Shade	Shade
Maireana		x	x	x	
Malacothamnus		x	x		
Malosma		x	x	x	
Malva		x	x	x	
Melaleuca		x	x		
Morella		x	x	x	
Myrsine		x	x	x	
Myrtus		x	x	x	
Oemleria	x			x	x
Olearia ×haastii		x	x	x	
Osmanthus		x	x	x	
Ozothamnus		x	x		
Peritoma		x	x	x	
Philadelphus	x	x	x	x	
Philotheca		x	x	x	
Phlomis		x	x		
Phylica		x	x		
Prunus	x	x	x	x	
Rhagodia		x	x	x	
Rhamnus		x	x	x	
Rhaphiolepis		x	x	x	
Rhus		x	x	x	
Ribes	x	x		x	x
Rosa	x	x	x	x	
Rosmarinus		x	x		
Rubus		x	x	x	x
Salvia		x	x		
Sambucus	x			x	x
Santolina		x	x	x	
Solanum		x	x	x	
Styrax	x		x	x	
Tagetes		x	x	x	
Tecoma		x	x	x	
Trichostema		x	x		
Vauquelinia		x	x	x	
Vitex	x	x	x		
Westringia		x	x		

PERENNIALS AND SUBSHRUBS

	Flowering Season				Exposure		
Genus	**Winter**	**Spring**	**Summer**	**Fall**	**Sun**	**Part Shade**	**Shade**
Acanthus		x	x		x	x	x
Achillea			x	x	x	x	
Agastache			x	x	x	x	
Allium		x			x	x	
Amaryllis				x	x	x	
Anigozanthos		x	x		x	x	
Armeria		x	x		x	x	
Asclepias			x		x	x	
Asteriscus		x	x		x		
Ballota			x		x		
Bergenia	x	x				x	x
Berlandiera		x	x		x		
Bidens			x		x		
Brodiaea		x			x	x	
Bulbine		x	x		x	x	
Bulbinella	x	x			x	x	
Calochortus		x			x	x	
Camassia		x			x	x	
Cistanthe		x	x	x	x	x	
Cyclamen	x	x		x	x	x	
Dichelostemma		x			x	x	
Dicliptera			x	x	x		
Diplacus		x	x		x	x	
Dryopteris	—	—	—	—		x	x
Dymondia			x		x	x	
Epilobium			x	x	x	x	
Epimedium		x				x	x
Erigeron		x	x		x		

	Flowering Season				Exposure		
Genus	Winter	Spring	Summer	Fall	Sun	Part Shade	Shade
Eriogonum		x	x		x	x	
Eriophyllum		x	x		x	x	
Erodium		x	x		x	x	
Eryngium			x		x		
Erysimum	x	x	x	x	x	x	
Eschscholzia		x	x		x		
Euphorbia		x			x	x	
Fragaria	x	x				x	x
Geranium		x	x			x	
Glandularia	x	x	x		x	x	
Glaucium		x	x	x	x		
Heuchera		x	x			x	x
Hunnemannia			x	x	x		
Iris	x	x			x	x	
Lavandula		x	x		x		
Leptosyne	x	x			x		
Lessingia		x	x		x	x	
Lupinus		x			x		
Marrubium		x			x		
Melampodium		x	x	x	x	x	
Monardella		x	x		x	x	
Narcissus		x			x		
Nepeta			x		x	x	
Origanum			x		x	x	
Pelargonium		x	x	x	x	x	
Penstemon		x	x		x	x	
Perityle		x	x		x	x	
Perovskia			x		x		
Phacelia		x	x		x	x	
Phlomis		x	x		x	x	

	Flowering Season				Exposure		
Genus	**Winter**	**Spring**	**Summer**	**Fall**	**Sun**	**Part Shade**	**Shade**
Phormium		x	x		x	x	
Polypodium	—	—	—	—		x	x
Polystichum	—	—	—	—		x	x
Ranunculus		x			x	x	
Rhodanthemum		x		x	x		
Romneya			x		x		
Salvia		x	x		x	x	
Sideritis		x	x		x		
Sisyrinchium		x			x	x	
Sphaeralcea		x	x		x		
Stachys		x	x		x	x	
Symphyotrichum			x		x	x	
Teucrium		x	x		x	x	
Thalictrum		x				x	x
Thymus		x	x		x	x	
Triteleia		x			x	x	
Verbena		x	x		x		
Wyethia		x			x	x	
Xanthorrhoea		x			x	x	

SUCCULENTS AND SUCCULENT PERENNIALS

	Flowering Season				Exposure		
Genus	**Winter**	**Spring**	**Summer**	**Fall**	**Sun**	**Part Shade**	**Shade**
Agave		x			x	x	
Aloe	x	x		x	x	x	
Beschorneria		x	x		x	x	
Bulbine		x	x		x	x	
Cotyledon			x		x	x	
Crassula	x	x	x	x	x	x	
Dasylirion			x		x		
Delosperma		x		x	x	x	
Dudleya		x	x		x	x	
Echeveria		x	x		x	x	
Furcraea	x			x	x		
Graptopetalum			x		x	x	
Hesperaloe			x		x	x	
Hesperoyucca		x			x		
Nolina		x			x		
Puya		x			x		
Sedum	x	x	x	x	x	x	
Sempervivum		x	x		x	x	
Yucca		x	x		x	x	

GRASSES AND GRASSLIKE PLANTS

	Active Growth Season		Exposure		
Genus	**Cool Season**	**Warm Season**	**Sun**	**Part Shade**	**Shade**
Andropogon		x	x	x	
Anemanthele	x		x	x	
Aristida		x	x	x	
Bouteloua		x	x	x	
Buchloe		x	x		
Calamagrostis	x			x	x
Carex	—	—		x	x
Elegia	—	—	x	x	
Elymus	x		x	x	
Festuca	x			x	x
Helictotrichon	x		x	x	
Juncus	—	—	x	x	x
Koeleria	x		x	x	
Lomandra	—	—	x	x	x
Melica	x		x	x	x
Muhlenbergia		x	x	x	
Schizachyrium		x	x	x	
Sporobolus		x	x		
Stipa	x		x	x	

VINES

Genus	Deciduous	Evergreen	Exposure: Sun	Part Shade	Shade
Aristolochia	x			x	
Bougainvillea		x	x		
Clematis	x			x	x
Hardenbergia		x	x	x	
Jasminum	x	x	x	x	
Lonicera	x	x	x	x	
Vitis	x		x	x	

ANNUALS

	Flowering Season				Exposure		
Genus	**Winter**	**Spring**	**Summer**	**Fall**	**Sun**	**Part Shade**	**Shade**
Clarkia		x	x		x	x	
Gilia		x			x	x	
Layia	x	x			x		
Limnanthes		x			x		
Lupinus		x			x	x	
Madia			x	x	x		
Mentzelia		x	x		x		
Papaver		x			x	x	
Phacelia		x			x	x	

Readings and Resources

California buckeye (*Aesculus californica*) is content in sun or part shade.

Readings

Bakker, Elna. *An Island Called California: An Ecological Introduction to Its Natural Communities*. Berkeley and Los Angeles: University of California Press, 1971.

Barbour, Michael, Bruce Pavlik, Frank Drysdale, and Susan Lindstrom. *California's Changing Landscapes: Diversity and Conservation of California Vegetation*. Sacramento: California Native Plant Society, 1993.

Bonine, Paul, and Amy Campion. *Gardening in the Pacific Northwest: The Complete Homeowner's Guide*. Portland, OR: Timber Press, 2017.

Bornstein, Carol, David Fross, and Bart O'Brien. *California Native Plants for the Garden*. Los Olivos, CA: Cachuma Press, 2005.

Christopher, Thomas, ed. *The New American Landscape: Leading Voices on the Future of Sustainable Gardening*. Portland, OR: Timber Press, 2011.

Dallman, Peter. *Plant Life in the World's Mediterranean Climates*. Berkeley and Los Angeles: University of California Press, 1998.

Darke, Rick, and Douglas Tallamy. *The Living Landscape: Designing for Beauty and Biodiversity in the Home Garden*. Portland, OR: Timber Press, 2014.

Faber, Phyllis, ed. *California's Wild Gardens: A Living Legacy*. Sacramento: California Native Plant Society, 1997.

Filippi, Olivier. *The Dry Gardening Handbook: Plants and Practices for a Changing Climate*. London: Thames & Hudson, 2008.

Filippi, Olivier. *Planting Design for Dry Gardens: Beautiful, Resilient Groundcovers for Terraces, Paved Areas, Gravel and Other Alternatives to the Lawn*. London: Filbert Press, 2016.

Francis, Mark, and Andreas Reiman. *The California Landscape Garden: Ecology, Culture, and Design*. Berkeley and Los Angeles: University of California Press, 1999.

Hall, Carol W., and Norman E. Hall. *Timber Press Guide to Gardening in the Pacific Northwest*. Portland, OR: Timber Press, 2008.

Harlow, Nora, ed. *Plants and Landscapes for Summer-Dry Climates of the San Francisco Bay Area*. Oakland, CA: East Bay Municipal Utility District, 2004.

Lowry, Judith Larner. *Gardening with a Wild Heart: Restoring California's Native Landscapes at Home*. Berkeley and Los Angeles: University of California Press, 1999.

Lowry, Judith Larner. *The Landscaping Ideas of Jays: A Natural History of the Backyard Restoration Garden*. Berkeley and Los Angeles: University of California Press, 2007.

Ogden, Scott, and Lauren Springer Ogden. *Plant-Driven Design: Creating Gardens That Honor Plants, Place, and Spirit*. Portland, OR: Timber Press, 2008.

Oudolf, Piet, and Noel Kingsbury. *Planting Design: Gardens in Time and Space*. Portland, OR: Timber Press, 2005.

Perry, Bob. *Landscape Plants for California Gardens: An Illustrated Reference of Plants for California Landscapes*. Claremont, CA: Land Design Publishing, 2010.

Rainer, Thomas, and Claudia West. *Planting in a Post-Wild World: Designing Plant Communities for Resilient Landscapes*. Portland, OR: Timber Press, 2015.

Reichard, Sarah H., and Peter White. "Horticulture as a pathway of invasive plant introductions in the United States." *BioScience*, 51 (2) February 2001: 103–13.

Robinson, William, and Rick Darke. *The Wild Garden: Expanded Edition*. Portland, OR: Timber Press, 2009.

Scheyer, J. M., and K. W. Hipple. *Urban Soil Primer*. U.S. Department of Agriculture, Natural Resources Conservation Service, National Soil Survey Center, Lincoln, NE: 2005.

Smith, Nevin. *Native Treasures: Gardening with the Plants of California*. Berkeley and Los Angeles: University of California Press, 2006.

Sunset editors. *The New Sunset Western Garden Book: The Ultimate Gardening Guide*. New York: Sunset Books, 2012.

Weaner, Larry, and Thomas Christopher. *Garden Revolution: How Our Landscapes Can Be a Source of Environmental Change*. Portland, OR: Timber Press, 2016.

Resources

Calflora: Information on wild California plants
https://www.calflora.org

California Center for Urban Horticulture, University of California, Davis, Water Use Classification of Landscape Species (WUCOLS IV)
https://ucanr.edu/sites/WUCOLS/

California Invasive Plant Council (Cal-IPC), The Cal-IPC Inventory
https://www.cal-ipc.org/plants/inventory

California Native Plant Society, Calscape: Restore Nature One Garden at a Time
https://www.calscape.org/

California Native Plant Society, Manual of California Vegetation
https://www.cnps.org/vegetation/manual-of-california-vegetation

E-Flora BC: Electronic Atlas of the Flora of British Columbia
http://ibis.geog.ubc.ca/biodiversity/eflora/

Great Plant Picks, an educational awards program of the Elisabeth Carey Miller Botanical Garden
www.greatplantpicks.org

Invasive Species Council of BC, British Columbia invasive plants list
https://bcinvasives.ca/invasive-species/identify/invasive-plants

Missouri Botanical Garden, Plant Finder
www.missouribotanicalgarden.org/plantfinder/plantfindersearch.aspx

Native Plant Society of Oregon
www.npsoregon.org

Oregon Building Codes Division, Oregon Smart Guide: Rainwater Harvesting
https://www.oregon.gov/bcd/Documents/brochures/3660.pdf

Oregon Department of Agriculture, Oregon Noxious Weed Profiles
https://www.oregon.gov/oda/programs/weeds/oregonnoxiousweeds/pages/aboutoregonweeds.aspx

Oregon State University, The Oregon Rain Garden Guide
https://seagrant.oregonstate.edu/sgpubs/oregon-rain-garden-guide

San Marcos Growers, Plant Index
https://www.smgrowers.com/plantindx.asp

National Gardening Association, USDA Hardiness Zone Finder
https://garden.org/nga/zipzone/

University of California, Davis, Arboretum and Public Garden, Arboretum All-Stars
https://arboretum.ucdavis.edu/arboretum-all-stars

University of California, Division of Agriculture and Natural Resources, Home Landscaping for Fire
https://anrcatalog.ucanr.edu/pdf/8228.pdf

Washington Native Plant Society, Plant Directory
https://www.wnps.org/native-plant-directory

Washington State Noxious Weed Control Board, The 3 Classes of Noxious Weeds
https://www.nwcb.wa.gov/classes-of-noxious-weeds

Xera Plants, Featured Plants catalog
https://xeraplants.com/plant-catalog

Index

D

E

F

G

H

I

J

K

L

M

N

O

Q

R

S

T

U

V

© Sheena Miraftabi

Nora Harlow is a landscape architect and longtime dirt gardener with wide-ranging experience in the San Francisco Bay Area, Los Angeles, and Davis, California. She was assistant editor of *Pacific Horticulture* magazine for many years and supervisor of water conservation for the East Bay Municipal Utility District in Oakland, where she designed low-water landscapes for District facilities. Among her many publications in far-flung fields are several on summer-dry gardening. She was principal author and editor of *Plants and Landscapes for Summer-Dry Climates of the San Francisco Bay Region*, co-editor of *The Pacific Horticulture Book of Western Gardening* (with George Waters), and co-editor of *Wild Lilies, Irises, and Grasses* (with Kristin Jakob). She has written numerous articles for *Pacific Horticulture* magazine and a bi-weekly column on summer-dry gardening that appeared in a dozen San Francisco Bay Region newspapers. Now officially retired, she is finally able to devote more attention to her own northern California garden, which by design will never be finished.

© Saxon Holt

Saxon Holt is a photojournalist who has spent more than 40 years exploring Kingdom Plantae and the fundamental importance of plants to the health of the planet. A lifelong gardener, he abandoned commercial photography when he discovered garden publishers and could no longer stay in a studio. His work has been featured in diverse publications, from *Architectural Digest* and *Pacific Horticulture* magazine, to *Smithsonian* and *Money* magazines. This is his 30th book. His work increasingly focuses on the relationship between gardens and the land, seeking a sustainable aesthetic that can enhance both gardener and the earth. He licenses his archive of garden photography at PhotoBotanic.com, where he publishes articles on plants and gardens and teaches garden photography. He is the photography program director at San Francisco Botanical Garden and is a fellow of GardenComm (Garden Communicators International).